# Education and Gendered Citizenship
## in Pakistan

## Palgrave Macmillan's
## Postcolonial Studies in Education

Studies utilizing the perspectives of postcolonial theory have become established and increasingly widespread in the last few decades. This series embraces and broadly employs the postcolonial approach. As a site of struggle, education has constituted a key vehicle for the "colonization of the mind." The "post" in postcolonialism is both temporal, in the sense of emphasizing the processes of decolonization, and analytical in the sense of probing and contesting the aftermath of colonialism and the imperialism which succeeded it, utilizing materialist and discourse analysis. Postcolonial theory is particularly apt for exploring the implications of educational colonialism, decolonization, experimentation, revisioning, contradiction, and ambiguity not only for the former colonies, but also for the former colonial powers. This series views education as an important vehicle for both the inculcation and unlearning of colonial ideologies. It complements the diversity that exists in postcolonial studies of political economy, literature, sociology, and the interdisciplinary domain of cultural studies. Education is here being viewed in its broadest contexts, and is not confined to institutionalized learning. The aim of this series is to identify and help establish new areas of educational inquiry in postcolonial studies.

### Series Editors:

**Peter Mayo** is Professor and Head of the Department of Education Studies at the University of Malta where he teaches in the areas of Sociology of Education and Adult Continuing Education, as well as in Comparative and International Education and Sociology more generally.

**Anne Hickling-Hudson** is Associate Professor of Education at Australia's Queensland University of Technology (QUT) where she specializes in cross-cultural and international education.

**Antonia Darder** is a Distinguished Professor of Educational Policy Studies and Latino/a Studies at the University of Illinois at Urbana-Champaign.

### Editorial Advisory Board

Carmel Borg (University of Malta)
John Baldacchino (Teachers College, Columbia University)
Jennifer Chan (University of British Columbia)
Christine Fox (University of Wollongong, Australia)
Zelia Gregoriou (University of Cyprus)
Leon Tikly (University of Bristol, UK)
Birgit Brock-Utne (Emeritus, University of Oslo, Norway)

### Titles:

*A New Social Contract in a Latin American Education Context*
Danilo Romeu Streck

*Critical Race, Feminism, and Education: A Social Justice Model*
Menah A.E. Pratt-Clarke

*Education and Gendered Citizenship in Pakistan*
M. Ayaz Naseem

# Education and Gendered Citizenship in Pakistan

*M. Ayaz Naseem*

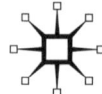

First published in 2010 by
PALGRAVE MACMILLAN®
in the United States—a division of St. Martin's Press LLC,
175 Fifth Avenue, New York, NY 10010.

Where this book is distributed in the UK, Europe and the rest of the world,
this is by Palgrave Macmillan, a division of Macmillan Publishers Limited,
registered in England, company number 785998, of Houndmills,
Basingstoke, Hampshire RG21 6XS.

Palgrave Macmillan is the global academic imprint of the above companies
and has companies and representatives throughout the world.

Palgrave® and Macmillan® are registered trademarks in the United States,
the United Kingdom, Europe and other countries.

ISBN: 978–0–230–61853–4

Library of Congress Cataloging-in-Publication Data

Naseem, M. Ayaz (Muhammad Ayaz), 1963–
        Education and gendered citizenship in Pakistan / M. Ayaz Naseem.
            p. cm.—(Postcolonial studies in education series)
        ISBN 978–0–230–61853–4 (hardback)
            1. Women—Education—Pakistan. 2. Sex discrimination in
    education—Pakistan. 3. Educational equalization—Pakistan.
        4. Women—Pakistan—Social conditions. I. Title.

LC2330.N37 2010
371.82209581—dc22                                              2010019022

A catalogue record of the book is available from the British Library.

Design by Newgen Imaging Systems (P) Ltd., Chennai, India.

First edition: December 2010

10 9 8 7 6 5 4 3 2 1

Printed in the United States of America.

I dedicate my work to
Dia
Who showed me
The power of the feminine

# Contents

# Series Editors' Foreword

In this groundbreaking study, M. Ayaz Naseem formulates and explores patterns germane to understanding complex systems of oppression at work in the subordination of women in Pakistan. His study carefully deconstructs how gender inequalities emerge in the broader society, especially within the context of political, religious, and public life. From there, he goes on to explore how educational discourses both reflect and influence the unequal gendering of citizenship that perpetuates the unequal and sometimes brutal treatment of women. A key concern here is the way gender discourses socialize boys and girls into accepting differently gendered identities or subjectivities, within the course of their everyday lives. Through his analysis, Naseem takes the position of being a moral ally in the feminist struggle for gender justice, arguing that as a male "insider," privileged both in his elite social location and in his gender, he is in a strategic position to contribute to the cause. His advocacy leads us to consider important questions tied to the dynamics of the role of male "moral allies" in feminist political change, as well as to consider more closely whether and how feminist philosophies may themselves be influenced when advocated by men. How do these factors contribute to changing unequally gendered power relations in all spheres of society?

As in prominent postcolonial studies, Naseem's research points to how important it is to study gender questions through exploring the broad social contexts that give life to the gendered order of society, particularly with respect to the "historically constructed patterns of power relations between men and women" (Connell 1987: 99) that shape subjectivities, policies, and practices. Although it is true that most societies have traditionally had a patriarchal gender order, that this book explores the gender order in Pakistan is of utmost importance to postcolonial theorists. In this country, male rulers, while not disallowing female education as is done by some of their more extreme counterparts, nevertheless subject women to deeply oppressive legal, religious, and customary regulations, including restricted movement outside the home, onerous dress codes, an inequitable location in the family, and cruel double standards in the sphere of sexual mores. This gendered repression can even still include practices of shaming or public stoning of women who have been raped, rather than directing punishment onto male perpetrators of such violence. Scholars working within "postfoundational" traditions of poststructuralist and postcolonial theory shift their focus from analyzing society in the context of narrow Eurocentric understandings of the state and social classes. Instead, as Naseem does here, they engage with a far more complex palette interweaving political, cultural, sociological, and psychological strands of analysis that can assist us to better understand societies that remain in the grip of these tensions.

Postcolonial perspectives require us to grapple with the tensions and contradictions at work between postcolonial visions of new nationhood and cultural independence, humanist moral imperatives of equity and equality, and the continuation of repressive colonial, traditional, and indigenous structures. Many modern societies, as Naseem well demonstrates, continue to show these tensions at work. In some of these societies, feminist struggles have successfully challenged the more repressive aspects of the patriarchal gender order, enabling women to win some basic rights including the vote, the legal principle of equal pay for equal work, a share of family property and child custody in cases of divorce, and entitlement to legal intervention against sexual harassment and abuse. Yet, gender inequalities are still stubbornly entrenched. There is probably no country in which women are equal or even nearly equal to men in the economy, workplaces, government, politics, and institutional religion. And, as becomes clear by this discussion of gender issues in Pakistan, the battle against the sexual abuse and exploitation of women still has a long way to go before it is won.

In concert with sexual oppression, postcolonial feminisms argue that in many decolonizing nations, the subordination of women is intricately connected to economics, patriarchy, politics, and education. The prevailing international economic structure maintains the poverty that puts both men and women in precarious economic circumstances, but patriarchal polities structure the societies in ways that cause particular adversity for women. Feminist scholars in postcolonial contexts often utilize analytical frameworks that study the overlapping of *practical gender issues* relating to the meeting of basic needs, and *strategic gender issues* relating to the power relationships between women and men, women and institutions of government, education, religion and the family, and the state and the international economy (Molyneux 1985; Hickling-Hudson 1999). These interconnections are underlined by the observation, "Democratic education from a feminist perspective involves a challenge to the social conditions that have sustained women's second-class citizenship and their experiences of poverty, violence, harassment and economic exploitation" (Marshall and Arnot 2008: 104). Naseem's study strongly reminds us that these interconnections are still very much alive and well in Pakistan.

The contentious area of women's sexuality repeatedly underlies the repression of the women whose experiences are documented by Naseem. He argues that the discourse in Pakistan has established double standards through different codes of gender behavior, which clearly assert men's superiority over women in sociocultural, economic, and political spheres. The misogyny that subordinates females to patriarchal cruelties is opposed by some Muslims, whatever their religious outlook, and upheld by others. This creates the sort of tension that makes change possible even in gender relations so set by tradition over time, so stable that most people do not question them, and so entrenched that we can call them a regime of gender oppression.

Naseem's study of gendered citizenship and education employs a poststructuralist methodology that he shows to be an effective tool for deconstructing this gendered regime. His sophisticated use of the Foucauldian approaches of genealogy show how Pakistan's gendered discourses socialize males and females into accepting certain types of gender identities that intersect with religious and political strictures and visions. The discourses of the polity, religious customs, and school textbooks of Pakistan are laid bare, allowing us to consider their origins and the extent to which bias and injustice underscore their content. Making such injustice visible is an important step in the struggle to transform the discourse, by teaching males and females that a different discourse is possible and desirable. We might add that it is important to highlight and deconstruct in this manner other gendered educational discourses, including the promotion of single-sex schools; skewed male-female staffing and leadership profiles; unjustly gendered divisions in curriculum design, texts, and materials; unjustly gendered pedagogical practices; and social relations that include sexual harassment and abuse. Such critical discourse analysis promotes the visibility that is a necessary step towards educational debate and political change.

There is no question that schooling is a key institution in Foucault's concept of the disciplinary society, so vividly utilized in Naseem's work to delineate gendered citizenship in Pakistan. The school is, as Taylor points out (2004: 90), "a key site for re-producing gender relations, but it is also a key site where change can occur"—a change that would prepare both boys and girls for pursuing the vision of equity and parity in their future social, cultural, political, and economic roles. More importantly, Ayaz Naseem's study on the gendering of education and citizenship in Pakistan challenges us to think more deeply about the ways in which we understand gender inequalities and schooling. His analysis explores how a society constructs gender inequalities to the detriment of women and the manner in which this process socializes consensus toward a taken-for-granted acceptance of this gendered perception of the world. Reading this text with the knowledge that gender divisions persist as severe global concerns encourages us to think about the particular ways in which not only Pakistan, but every society, keeps women at the margins of political and economic power.

ANNE HICKLING-HUDSON, ANTONIA DARDER,<br>AND PETER MAYO

# References

Caldwell, C. (2009) *Reflections on the Revolution in Europe. Immigration, Islam and the West.* New York: Doubleday.

Connell, R.W. (1987) *Gender and Power.* Sydney: Allen & Unwin.

Hickling-Hudson, A. (1999) "Experiments in Political Literacy: Caribbean Women and Feminist Education." *Journal of Education and Development in the Caribbean* 3 (1): 19–44.

Marshall, H., and M. Arnot (2008) "The Gender Agenda: The Limits and Possibilities of Global and National Citizenship Education." In Joseph Zajda, Lynn Davies and Suzanne Majhanovich, (Eds.), *Comparative and Global Pedagogies. Equity, Access and Democracy in Education.* Netherlands: Springer, pp. 103–24.

Molyneux, M. (1985) "Mobilisation Without Emancipation? Women's Interests, State and Revolution in Nicaragua." *Feminist Studies* 11 (2): 227–54.

Taylor, S. (2004) "Gender Equity and Education: What Are the Issues Now?' In Bruce Burnett, Daphne Meadmore and Gordon Tait (Eds.), *New Questions for Contemporary Teachers.* French's Forest, NSW: Pearson Education Australia, pp. 87–100.

# Acknowledgments

Although the sins of omission and commission are solely mine, this research is a collaborative effort of many individuals and organizations. I may not be able to thank them all or enough but thank them I must. Some individuals who helped me immensely wish to remain anonymous and I respect their wishes and thank them for all they did for me.

First, I wish to acknowledge the contribution of those who participated in my research project, the teachers and students of two schools who shall remain unnamed. They in their own ways opened my eyes to dimensions and dynamics of curricula that would never have crossed my mind. They gave me their time, support, and above all love, and without their presence and support this research would not have been possible.

Gratitude is due to numerous scholars in Pakistan who took out time to talk to me during the fieldwork. I wish to acknowledge Dr. Pervez Hoodbhoy of Quaid I Azam University; Dr. Abdul Hamid Nayyar, Dr. Saba Gul Khattak, and Dr. Lubna Nazir of SDPI; Dr. Haroona Jatoi, Joint Educational Advisor, Curriculum Wing, Ministry of Education; Dr. Fauzia Saleemi, Director of the Punjab Textbook Board, Lahore; Dr. Sarfraz Khawaja Academy of Higher Education and Planning, Islamabad; and various officials of the Ministry of Education, Islamabad, who wish to remain anonymous. My thanks also go to various officials of the Punjab Textbook Board, textbook writers, and subject specialists who spoke to me and whose real identities I cannot reveal.

The staff at various libraries in Pakistan especially the National Library of Pakistan; Academy of Higher Education and Planning, Islamabad; Women's Division, Islamabad; the Quaid I Azam Library, Lahore; and the Punjab Textbook Board, Lahore, helped me immensely in locating and xeroxing materials. My heartfelt gratitude goes out to all of them. However, I particularly wish to acknowledge the support of Mr. Aurengzeb Malik, librarian, Punjab Textbook Board, for his enthusiastic help in locating the historical materials. I must say that Mr. Malik's knowledge of textbook publishing in Pakistan is miraculous. Without him, locating data would not have been extremely difficult.

I also wish to acknowledge my friends, colleagues, and especially my students at the Department of Education, Concordia University, Montreal. Each one of them in their own way always supported my research. I would especially like to thank my doctoral student and research assistant, Ms. Amna Zuberi, who helped me with the formatting and word processing issues. A big thanks is also due to Kamran Sheikh for his support and perpetual humor.

I would also like to thank Julia Cohen, who originally encouraged me to publish this work and who commissioned the contract. I would also like to thank Samantha Hasey, who took over from Julia and who was with me patiently and sustainingly through out the project.

My gratitude also goes out to the series editors of the Palgrave-Macmillan Postcolonial Studies in Education Series for their encouragement and more-than-helpful suggestions in the initial review. Special thanks to the anonymous reviewer whose comments and suggestion helped me immensely to refine the conceptual arguments of the work.

A longer version of chapter 5 appeared in the *International Journal of Inclusive Education* 10 (4–5): 449–67 (http://www.informaworld.com). I would like to acknowledge Taylor and Francis Publishers in granting the permission to reuse some parts of the original paper.

I am indebted to my family, especially Ayisha and Zahoor and my nephews Ahmad and Mohammad Shahnawaz, whose love and unending support I could always count on.

Finally, the person without whom nothing is and nothing would have been possible. My wife, my critic, my collaborator, my sounding board, secretary, cook, and lover—Dia. She understood and shared my commitment and always pulled me up when I was down, kept me human and sane, and was with me every step of the way.

# Urdu and Arabic Words

| | |
|---|---|
| Ailan Farmaya | Announced |
| Alim | Scholar |
| Arz e Watan | Country |
| Ayyar | Cunning |
| Bad Kirdar | Of loose morals |
| Bahadar | Brave |
| Chaddar | Veil |
| Chardewari | Four walls |
| Chiragh e Khana | Domestic light |
| Deeni Madaris | Religious schools |
| Dharti | Land |
| Dharti ki Kokh | Land's womb |
| Dhoban | Laundry woman |
| Dubai Chalo | Let's go to Dubai |
| Dupatta | Thin veil |
| Dushman | Enemy |
| Dusman Hawabaz | Enemy pilot |
| Ghasiara | Grass cutter |
| Ghazi | Victorious soldier |
| Ghazva/Ghazvat | Battle/Battles |
| Haq Mehar | An amount owed to wife at marriage |
| Hazrat | Revered |
| Hijra | Migration |
| Hudd | Law |
| Hudood | Laws |
| Islam ka Qilla | Castle of Islam |
| Ismaili | Shi'ite subsect |
| Jihad | Holy war |
| Kafir | Infidel |
| Kalima e Tauheed | Affirmation of the oneness of God (First of the five main pillars of Islam) |

| | |
|---|---|
| Kameez | Tunic |
| Khidmat | To serve |
| Kisan | Farmer |
| Lakarhara | Woodcutter |
| Lazmi | Compulsory |
| Madr e Millat | Mother of the nation |
| Madrassah | Religious school |
| Majazi Khuda | God in this world |
| Makkar | Devious |
| Mandi Merchants | Grain merchants |
| Mandir | Temple |
| Masjid | Mosque |
| Memon | Muslim subsect |
| Meri Kitab | Primer |
| Minar e Pakistan | Pakistan Monument |
| Mohallah/Muhalla | Neighborhood |
| Mouazzin | One who calls for prayer |
| Muasharti Uloom | Social studies |
| Mustaqil | Permanent |
| Mutala e Pakistan | Pakistan Studies |
| Naib Nazim | Deputy Administrator |
| Nazim | Administrator |
| Nishan e Haider | Highest military award in Pakistan |
| Nizam e Mustafa | Prophet Muhammad's System |
| Pir | Spiritual counselor |
| Purdah | Observance of veil |
| Qadiani | A sect that does not believe in the finality of Prophet Muhammad's prophethood |
| Qanun e Shahdat | Law of evidence |
| Qaum | Nation |
| Qisis and Diyat | Blood money |
| Quaid | Leader |
| Quaid i Azam | Great leader (a title of M. A. Jinnah) |
| Rahnatullah Alaihi | May he/she be blessed |
| Raja | King |
| Randi | Prostitute |
| Sati | Widow immolation |
| Shaheed | Martyr |
| Shalwar | Loose trouser worn in South Asia |
| Shama e Mehfil | Party light (socialite) |
| Shariat | Islamic law |

| | |
|---|---|
| Shia | A subsect in Islam |
| Sub kuch | Everything |
| Taluqdar | Land officer |
| Tanzeem | Organization |
| Tehsildar | District officer |
| Ummah | The entire Muslim fraternity |
| Ummatul Momineen | Mother of the true believers |
| Ustad | Teacher |
| Ustani | Teacher (female) |
| Virasat | Inheritance |
| Wahisha | Of loose moral character |
| Watan | Country |
| Zameen | Land |
| Zina | Fornication |
| Zina Bil Jabr | Rape |

# Chapter 1

# Contextualizing Articulations of Women in Pakistan

Since its inception as an independent nation state in 1947, Pakistan has come a long way in many respects. Over the years it has developed a number of key economic and social sectors and has created a number of institutions from scratch. While rapid development and/or improvement is apparent in areas such as industry, finance, media, and defense, the same cannot be said for education. This sector has seen few gains and many losses. The latter can be gauged in terms of relatively low rates of investment in education, falling enrollment rates, high drop-out ratios, inadequate teacher training, lopsided gender balance, lack of political will and patronage, and politically motivated agendas guiding curricular design and development. While successive governments have vowed to address the ills of the educational system and have boasted about the increase in literacy rates, the situation on the ground does not support either the promises or the tall claims.

While on paper the literacy rates might have gone up two- to threefold (largely due to various definitions of "literacy"), this has not resulted in any significant or meaningful empowerment for the people of Pakistan. Females have especially found themselves on the losing side of the equation. Not only have they been excluded by and large from the system in terms of access to education (enrollment, access to and choice of subject areas, etc.) but more importantly, those who have managed to get into the system have been marginalized by the educational discourse, especially by the curricula on offer at government schools where the majority of students enroll. The educational discourse in Pakistan through the curricula imparts education that constitutes particular gendered subjectivities that exclude and deprivilege some (women, minorities, and marginalized groups) while including and privileging a select few (such as men, military personnel, and upper strata of the society).

The status of women in Pakistan in the last two decades presents an intriguing picture. On the one hand, women have been subjected to the worst kinds of social, political, economic, and juridical discrimination, while on the other, this period has seen an upsurge of spirited resistance by Pakistani women to such injustices. Since the 1980s, educational, media, and judicial discourses (among others) have directly and indirectly constituted gendered subjectivities that accord unequal citizenship status and rights to women and men. The positioning of gendered subjectivities by these discourses has not only exacerbated and intensified the inequalities between men and women; it has also brought about a major alteration in gender relations and changes in the way men and women relate to the state and society in Pakistan.

A good exposition of these changes is apparent in the subject positioning in the judicial discourse in Pakistan. Take for instance the case of *zina* (literally, "fornication") in the *hudood* (a legal ordinance supposedly based on the juridico-legal edicts of Islamic jurisprudence). This ordinance, which was made a part of the legal system in Pakistan in 1979 by a decree of the military dictator General Zia ul Haq. prescribes the "Islamic" punishment of death by stoning for fornication outside wedlock. The prescribed punishment can only be awarded should there be four "pious" male witnesses who can vouch that such an act was committed. While the law applies to both men and women, in its application it has proved to be more detrimental for women than men as more women than men have been incarcerated under this law.

The situation grows even murkier when it comes to cases of *zina bil jabr* (rape). Due to the unequal and unjust status of women that emanates from the subject positioning within the discourse, the onus for proving the rape is on the woman. In other words, if and when a woman in Pakistan reports a rape case, it is up to her to prove that such a crime was committed by producing four "pious" male material witnesses. In cases where women have not been able to do so, not only has the perpetrator of the crime been absolved of the crime, the victim of the rape has been charged with seducing the man to fornicate. She is thus liable to be punished by stoning to death under the *hudood* ordinance.

Furthermore, the laws of *shahdat* (evidence), *virasat* (inheritance), and *qisas* and *diyat* (blood money) stipulate that women's evidence, inheritance, and blood money (in the event she is murdered) will be worth only half that of men. The discourse constitutes the subjectivity of women in such a way that not only are practical matters relating to the legal, social, and economic status of women affected, but more importantly the very meaning of "woman" is fixed as half that of "man" in all social practices as well. In other words, the sign "woman" is "momentized" as half of man. While the juridico-legal position of "woman" with respect to citizenship is still that of equality between men and women, in practical terms her citizenship status is also reduced to half that of man.

The media discourse in Pakistan (print and electronic) similarly has fixed the meanings for "woman" in Pakistan. This fixation of meaning is partly a result of the process of "othering" (creation of the "other"). The woman in media discourse is constituted as the "other" of the man, who is articulated as the "national hero," the guardian of national and ideological frontiers, the provider of subsistence, and so on. In each case a superior meaning is ascribed to the "other" of the woman, thus relegating her to the level of an inferior citizen. Furthermore, the discourse of sports draws upon the military discourse to present sportsmen as heroes who safeguard the honor of the motherland. Political signs relating to social justice are relegated if not excluded from the discourse and are replaced by the laurels of national heroes, military, and sports, and are invariably male. On the other hand, for a long period of time (1977–1985), female sports were banned, women's teams were withdrawn from international competitions, and women were disallowed to play any spectator sport. In other words, women were excluded from the meaning of "national hero," which could only be understood in male (and militarist) ways.

Such fixation of meanings is also evident in androcentric films and television plays that reify the notions of *chadar* and *chardewari* (the veil and the four walls of the home). The woman outside such "moments"[1] is considered deviant and thus liable to be disciplined by any means. This is particularly evident in cases of violence against women. Take for example the rape case of Veena Hayat. A popular socialite and scion of a political family from northern Pakistan, Veena was allegedly gang raped by the son-in-law of the then president of Pakistan, Ghulam Ishaq Khan, and his cohorts (*Dawn*, November 27, 1991). The media discourse portrayed Veena as a debauched socialite who consumed alcohol and lived to party. In other words, the press and large sections of the society justified the heinous act of Veena's rape because she did not adhere to the meaning of a "good," "respectable" woman that was momentized and articulated by the discourse. The subtext of the media discourse was that she was inviting the act. The perpetrators of the crime were never punished. Within the media discourse, we find processes through which fixation of meaning becomes so conventionalized that it has come to be seen as natural.

The educational discourse in Pakistan is perhaps the most significant site of discursive fixation of meaning and the constitution of gendered subjectivity. Drawing upon the interdiscursivity of other discourses such as those of law, media, state, and military, the educational discourse is not only the primary site where meanings of signs such as woman, man, mother, father, and so on are gendered; it also provides the techniques of discipline (the school system and pedagogical practices) and surveillance (examination) for naturalization of meaning (the process through which meanings of different words/terms/notions are rendered unproblematic) and normalization of subjects (the process through which subjects stop questioning).

Let me take an example from school curricula in Pakistan to illustrate this point. A major theme that runs through the social studies and history curricula, from grades 1 through 8, relates to the place of women in the social system of the country. Women are portrayed as mothers, sisters, and wives whose primary duties include childbearing, childrearing, and other household responsibilities. Their virtues include sacrifice, devotion, religiosity, and dedication to family. The curricula invoke a sharp public-private dichotomy. Women, according to this dichotomy, exist and operate exclusively in the private domain and totally outside the "public" domains of politics and economics. Even their roles as educators are confined to the four walls of the "home." The educational discourse, through curricula and textbooks, fixes the meaning of what it means to be a woman. It thus in part constitutes the subjectivity of the female subject.

It is in the context of the constitution of gendered subjectivity that I question in this study the unproblematic use of the dictum of development discourse that education results in empowerment for women. I posit that the kind of education that is being imparted in Pakistan actually disempowers large sections of the society, especially women and the marginalized groups. This is not to say that women should withdraw their demands for access to education. Rather, the working of the educational discourse including its technologies through which women are disempowered must be critically examined and understood so as to make education meaningful and empowering to the lives of women in Pakistan.

Contrary to some popular imagery, women in Pakistan have neither been mere bystanders nor merely victims of the discourses that constituted their subjectivities. In recent years, women in Pakistan have engaged with and challenged the marginalizing discourses and discursive practices by means of agency, activism, and scholarship. The Women's Action Forum (WAF), for example, was one of the few strata of the society that continuously and incessantly challenged the marginalizing discourses during the dark days of the martial law from 1977 to 1988. Similarly, feminist scholars and activists located both within and outside Pakistan critically examined, analyzed, and challenged the state of women in Pakistan from multiple epistemic and theoretical positions.[2]

This engagement and the scholarship thus produced is indeed perceptive and articulate. It provides extremely important insights into the ways and processes through which the notion of "woman" has come to be articulated in various discourses in Pakistan. It offers both empirical and theoretical depth to an understanding of how women have come to be represented the way they are in Pakistan. My purpose in writing this book is neither to examine and fill the empirical gaps in this literature nor to critique the epistemological stance of the analyses that this body of literature offers. From a position of shared political and ideological commitment, my focus, in reading, rereading, and drawing upon the rich body of literature on women in Pakistan, is on trying to reconceptualize the questions rather than on problematizing the conclusions reached by these studies.

In this respect I take two issues that need a critical revisitation. First, almost all of these studies take a unitary notion of state as the level of analysis. This statist orientation of literature on gender relations in Pakistan can partly be attributed to the influence of Western feminist theorizing on Pakistani feminist scholarship and partly to the fact that much of this scholarship was produced at a historical moment when the women's movement in Pakistan was evolving (especially with respect to the despotic military regime of General Zia ul Haq). Here I am referring to most of the literature produced since the late 1970s, and by no means do I imply that no feminist literature was produced prior to this period.[3]

These historical moments invoked different epistemological stances that led to different questions being raised. The general focus of this body of literature on women in Pakistan was on questions related to the relationship between women and the state and how this relationship "constructed" identities. Such a line of questioning produced analyses that took a rather static (and unitary) view of both the state and the identities that were constructed in the relationship. It led to a conceptualizing of power as concentrated in very few places (state, elite, etc.), and as top down and repressive. Such a framing of questions and issues led to essentializing the notion of woman as victim and power as repressive. Theorizing from this position was thus more focused on questions of "why," that is, why power was being exercised (legitimacy of state or regime, perpetuation of patriarchy) rather than on questions of "how," that is, how power is/was exercised (processes, techniques, strategies) and how subjectivities are constituted.

In other words, the focus of the abovementioned analyses has been more on power than power relations, more on gender than gender relations. Looking at the constitution of gendered subjectivities (construction of identities) through the lens of the state-gender relationship obfuscates the fact that subjectivities are primarily a discursive product of gender relations. The relationship between subjects of the discourse constitutes subjectivities. As a result, a number of issues have remained unexplored or underexplored or have simply been brushed under the carpet because of the lack of conceptual tools to analyze them.

A second issue is that of examination of gender relations in Pakistan along the secular-religious binary. Inherent in such analyses is the assumption that the former tendency has been represented by the nationalist, postindependence state and that it has always contested the proponents of the religious order in Pakistan. Also inherent in this line of argument is that the former is/was "better" for the women of Pakistan than the latter (see for instance Weiss 1988; Esposito 1994; Alavi 1986; Zafar 1988). This line of argument is based on the assumption that, once the postindependence nationalist state succeeds in displacing (relatively) the religious political and ideological agenda, repression of women will automatically disappear. However, empirical evidence, not only from Pakistan but also from other Muslim countries (like Turkey and Egypt), suggests that such is not the case. The repression of women has continued even in cases where the modernizing national state has displaced the religio-political agenda. My argument is that the modernizing nationalist state and the religio-political agenda are not in a dichotomous binary relationship. Rather, they are produced discursively by the same discourse of modernity in postcolonial societies. Thus strategies of resistance also have to be located within the same discursive practices.

In taking issue with these orientations, I hope to shift the focus from finding allies in one structure or another to finding points of resistance within the discourse that constitutes subjects and subjectivities and that, in turn is constituted by them. Since modernity and religious obscurantism in Pakistan are grounded in particular historical and local discursive contexts, therefore, by retheorizing the relationship between women and the state in Pakistan, I aim to reach a more historicized and locally situated explanation of gender relations in Pakistan. An historically and locally situated explanation avoids invoking an essentializing and universal notion of woman that characterizes certain Western feminist as well as elitist feminist accounts about women in Pakistan.

A second area that this book addresses is education in general, and women and education in Pakistan in particular. As compared to the body of literature on issues related to women in Pakistan, that which deals with issues related to women and education in Pakistan presents a gloomy picture. I make this claim on two bases. First, in the sixty years of Pakistan as an independent, national entity, the amount of literature produced on issues related to education is downright miniscule. Though quantity is never and should never be taken as an indicator of the quality of scholarship, it does serve to show how little thought seems to have gone into research on such an important topic. Second, research on issues related to education in Pakistan has been done from an extremely narrow range of epistemic positions, which serves to show the limited perspective of the debate on educational issues in Pakistan.

There are only a handful of studies available on this important topic. Discounting the monographs on the particular technical aspects of education in Pakistan, there are no more than a handful of book-length studies dealing with the state of education in Pakistan (Qureshi 1975; Quddus 1979; Saigol 1993, 1995), few studies in comparative perspective (Saqib 1983; Curle 1973), country studies by intergovernmental organizations such as UNESCO (2002), studies that focus on madrassah education (Rahman 2004; Riaz 2008; Fair 2008), reports (Nayyar and Salim n.d.), and a few edited volumes (Hoodbhoy 1998; Edgar and Lyon 2009).

The situation with respect to women and education is even worse. There exists only one book-length research study on this topic (Saigol 1995) and a handful of research articles. Of the limited number of studies available, only a few (Saigol 1995; Aziz 1993; Hasnain and Nayyar 1997; Hoodbhoy and Nayyar 1985) touch upon the relationship between education (in general) and curricula and text-books (in particular) and empowerment. Only Saigol (1995) and to a limited extent Nayyar and Salim (n.d.) have looked at the articulation of gender in educational discourse. Thus, I hope to contribute to this literature in the hope of providing intellectual stimulus to further research in this respect.

As mentioned earlier, most of the studies that have been done deal with the problems of education in Pakistan in general and with issues related to women and education in Pakistan in particular from a very narrow epistemic base. Thus, while these studies provide figures useful for understanding some of the apparent problems plaguing the system, such as low enrollment rates, high drop-out rates, teachers' training, and structural problems, they either gloss over issues such as the relationship between education and empowerment, the construction of identities, and the quality of education, or with the help of macroeconomic statistics paint a rosy picture that is nowhere near the real situation on the ground.

## Organization of the Book

In examining how gendered subjectivities are constituted in and by the textbooks that are used in the government-run public schools that the majority of students (both male and female) in Pakistan attend, I look at the textbooks that curricula prescribe for Social Studies (for classes IV to VIII) and Urdu (for classes I to VIII). I particularly examine textbooks published under the 2002 and 2004 curricular reforms. There are five textbook boards in Pakistan (one for each province and one recently established textbook board for the federal capital and the educational institutions run by it). I chose to look at the textbooks produced by the Punjab Textbook Board (PTB) and the Federal Textbook Board (FTB). While I do examine the textbooks of the other three boards—namely, the Baluchistan, North West Frontier Province (NWFP), and Sind textbook boards—in order to draw comparisons, I concentrate more on the PTB and FTB. Furthermore, there is not much difference between the PTB/FTB textbooks and those produced by other textbook boards in terms of content. Thus, the corpus I chose is by and large representative of the whole of Pakistan.

The curricular advice that guides the production of textbooks by any of the boards comes from the Curriculum Wing of the federal Ministry of Education in Islamabad. This curriculum advice also forms a part of my data corpus. It should be noted that I look only at those textbooks that have been produced in the last ten to twelve years. There are two reasons for this: first, even the textbook board libraries do not have full records of the textbooks published by them; and second, the last twelve years have seen major curricular changes that promised to provide insights into whether there is a process of change underway.

There are two further delimitations that I apply to the present study. First, while I focus primarily on two discourses, namely, the educational discourse and that of the state and citizenship in Pakistan, this by no means implies that other discourses are unimportant or do not play any considerable part in the construction of identities. Other discourses, such as those of defense, media, and development, to name a few, do play a significant role in this respect. However, to incorporate all of these to the same degree would have meant extending the limits and scope of my research. I thus look at them in the context of interdiscursivity, i.e., the ways in which the educational discourse in Pakistan draws upon these discourses to constitute subjectivities. Second, while I have focused on how female subjectivities are constituted by the discourse under study, this does not mean that these are the only identities constructed in the process, or that other gendered identities are of any lesser importance. In fact, I have touched upon the identity construction of masculine identities as well, but to a lesser extent.

I have organized the book into eight chapters. Chapter 2 examines the following questions: Who are the subjects of educational research? How is she positioned in the discourse especially in relation to power, knowledge, and the state? What space does this subject of educational inquiry occupy? And how is her subjectivity and agency constituted within and by the educational discourses? This chapter sets the conceptual parameters of the discussion to follow (in other chapters) by reviewing, critiquing, and engaging with poststructuralist theory.

Chapter 3 provides an overview of Pakistan's educational system and the policy discourse. In this respect the focus is on the numerical dimensions, infrastructure, educational policy making, and curriculum development in Pakistan.

Chapter 4 makes a case that women in Pakistan are primarily understood through the meanings fixation by competing and contesting discourses of the state, media, law, and education, among others. To this end, an effort is made to historicize women in relation to the state in Pakistan by juxtaposing the discourses of the state and gender in Pakistan in a way that discourse of gender provides insights into the evolution and working of the discourse of the Pakistani state.

This chapter also examines how other discourses such as those of law and media contest to articulate how women in Pakistan are understood by themselves and by others. The main aim of the chapter is not to provide a historical context of these notions but to trace the contours of the discourses of gender relations and the state through the Foucauldian strategy of archaeology, i.e., how these discourses came about. The chapter also briefly traces a genealogy of these discourses to uncover the processes of inclusion and exclusion, fixation of meaning (moments), creation of a body of floating signs (i.e., elements whose meanings have not been fixed), and how these elements mount resistance to the fixation of meaning of signs.

Chapters 5 and 6 situates the subjectivity constitution in Pakistan's educational discourse along two dimensions, namely, the nodal points (signs or concepts around which the meaning fixation of other signs or concepts is clustered) and the moments (fixing of meanings for particular signs). The main argument presented in these chapter is that in Pakistan, constitution of gendered subjectivities in the educational discourse is ordered around the nodal points of nationalism and Islam. Coding categories for "moments" include normalization (definition of normal by the text), exclusion (definition of difference by the text), totalizing (specification of totality as means of homogenizing explanation/theory), classification (differentiation of groups), space (demarcation of personal, physical, and psychological space), time (division of historical and present time), and self (subject's view of identity). The chapters also examines how gendered subjectivities (identities) are discursively constituted in the educational discourse in Pakistan. This is followed up by an analysis of the consequences of such subjectivity constitution for citizenship across genders and marginality.

In chapter 7 I examine technologies of classification, normalization, in order to understand the Construction of the "other" in Pakistan's educational discourse. In chapter 8, I draw conclusions from the discussion in the preceding chapters.

## Voice and Signature

Let me end the introduction by situating my voice and signature, in other words, my situationality in the context of this book. To begin with, I am a male. I come from a predominantly Muslim, extremely patriarchal, and frequently nondemocratic society. In my society, patriotism, nationalism, and ethnic, religious, and gender chauvinism and even obscurantism enjoy a premium position supported and nurtured by the state and Pakistani society alike.

At the same time, I come from an upper-upper-middle-class, predominantly technocrat-professional, urban, educated family where the chauvinisms of the kind outlined above run subterraneously. My family background provided me with the added advantage of going to schools where the medium of instruction was English. This education privileged me in the context of Pakistani society and also afforded me the opportunity for graduate studies and travel abroad. I am thus also familiar with various Western and Eastern texts and cultural sources that kindle in me the desire to be a liberal, a democrat, and a feminist. On the other hand, this very education makes me a cultural and national patriot following Anthony Kwame Appiah, Edward Said, Frantz Fanon, and Chinua Achebe. Postcolonial patriotism still occupies its tiny niche in me and in moments of frustration, I still (even if erroneously) blame the West and colonialism for everything that goes wrong in my life.

These tensions thus pull me in at least three different directions. One, being a postcolonial citizen, I am located on the margins of the global polity and economy although certainly not in the margins of the global society. I am thus a disadvantaged subaltern and am marginal in that respect. On the other hand, I am a middle-class male, which makes me a cohort if not an actual exploiter, a centrist or a power holder per se. I am thus at the center and at the margins at the same time. I am both the oppressor and the oppressed. My dilemma is twofold. First, like Uma Narayan (1989), I do not want to be misunderstood. By pointing out what is wrong with my society, I do not want to convey an impression that any other society is better than or superior to it. Similarly, I do not want to be thought of as a self-loathing Pakistani. Second, being a feminist, I want to be accepted as one, something that most women will be suspicious of, just as black feminists were not too comfortable with cross-ethnic feminist scholarship (read: white feminist scholarship) (hooks 1988).

These dilemmas do present me with problems with respect to my position within the inquiry, which certainly reflect on *my* narrative of the constitution of gendered subjectivities in Pakistan. They also point to the position where I can see myself within the research. They point to my position as an insider. This is the position that colors and guides my narrative and gives me an advantage in accessing and interpreting certain knowledge that might not be accessible to others. I thus use the insider's advantage to gather data on how discursive practices in education in Pakistan construct gendered subjectivities, both male and female. My position as an insider helps me make sense of the narratives of the subjects who are exposed to curricula and texts that aim at engendering identities. For example, as a Muslim, upper-class, educated male I have access to knowledge and information on how oppression works that might not be available to females or members of minority and subaltern groups. I can use my insider's advantage to the benefit of these groups by bringing this information out in the open.

## A Note on Conventions

There are a number of conventions used in this book that are different from those used traditionally. First, I use the first person singular throughout the text. Such use is borne out by my conviction that I am, as a researcher, an integral part of the research process and in the end it is my interpretation and experience of the discourses that I am explaining, along with those of other subjects of the discourse that are being presented to the reader. Second, while referring to the researcher conceptually, I use "she/her" instead of the traditional "he/him" or the more politically correct "they." This usage is partly grounded in my feminist intellectual orientation and partly in the fact that I am a feminist activist wishing to engender the referent with feminine attributes and thus move away from the "Enlightenment" referent who is inevitably male.

Third, informed by poststructuralist theorization, I will use "subject" instead of "man," "subjectivity" instead of "identity," and "subjectivity constitution" instead of "identity formation/construction." Fourth, terms such as post-*modern*, post-*structuralist*, and post-*colonial* are written both with hyphens and without hyphens. Some scholars (Hoskins 1990) conceptualize the hyphen as representative of a certain space. For instance, Hoskins writes that "what we may need to consider is how "the educational" may in different epochs, in different ways, function as the hyphen in the power–knowledge relationship" (1990: 51). Other scholars such as Hickling-Hudson argue that the hyphenated versions of the terms should refer to the temporal context, and the nonhyphenated version to the theoretical approach. I follow such articulation and use the nonhyphenated terms.

# Chapter 2

## Michel Foucault and I: Applying Poststructuralism to the Constitution of Gendered Subjects in Pakistan's Educational Discourse

One of the big questions before me is how useful is it to use poststructuralism, a primarily Western (and to some a canonical) theoretical lens to examine and understand gendered subject and subjectivity constitution in a specifically non-Western society like Pakistan. A second issue before me is how problematic is my position as a male in trying to understand gendered subjects in relation to Pakistan's educational discourse.

In order to address these issues, I engage with some of the methods and tenets of poststructuralism in this chapter as they come to me via Michel Foucault. In this respect I engage with the genealogical method and the notions of power, power/knowledge, and space as power in order to see the extent of their analytical prowess in understanding gender in relation to education in Pakistan. I also interrogate the interrelated notions of subject, subjectivity, and agency as they come to me via Foucault. My intent is to establish if these notions that were developed in a primarily Western context are at all relevant to an analysis of gender and education in Pakistan. Of particular importance to me are Foucault's rearticulations with respect the state and governmentality and their relevance to the study of gender dynamics in a formerly colonized state. Finally, I problematize myself as a researcher in relation to issues of voice. In doing so, I lay down the groundwork for a position from where male feminists acting as moral allies can contribute to feminist scholarship on a number of issues, including but not limited to the constitution of gendered subjects and subjectivities in relation to educational discourses.

Education as a field was perhaps late in looking at the possibilities of using poststructuralism for analyzing and addressing educational issues. However, in the last two decades several scholars have turned to this tradition in order to find answers to the pedagogical questions of our times.[1]

From a Foucauldian perspective, there are three main functions of education in modern societies: the generation of discourses, normalization[2] through discipline[3] (and the resultant constitution of subjects and subjectivities), and gaining control over bodies (Dreyfus and Rabinow 1983: 67). Education is also inherently related to Foucault's notion of power/knowledge. It is at the site of education that some voices are excluded while others are legitimized and validated. In other words, education is that site where historical and specific discourses are generated. For Foucault, "every educational system is a political means of maintaining or modifying the appropriateness of discourses with knowledge and power they bring with them." Furthermore, educational discourses through pedagogical methods, such as school discipline, make "children's bodies the object of highly complex systems of manipulation and conditioning" (Foucault 1971: 46, cited in Dreyfus and Rabinow 1983: 67).

Education, like any other social process, uses techniques to legitimize (or normalize) institutions of society. Furthermore, enmeshed with economics, politics, history, and culture, education constructs and shapes human beings as subjects (Ball 1990). As Foucault notes in the context of Europe, schools as institutions/apparatuses of discipline were experienced by far greater populations than ever before. Using what he terms "dividing practices," schools constructed identities through techniques of organization such as criteria of admissions and evaluation (tests, exams), separate and different curricula and pedagogies, and forms of student-teacher relationships. Identities and subjectivities thus constructed are carried forward by the students into the larger society (Ball 1990: 4). By defining what is normal/subnormal, subjectivities are created, stigmatized, and thus normalized.

What apparently emerges from the preceding discussion is indeed a very bleak picture. It appears as if education is merely a tool, an instrument of power used to suppress, subjugate, and control. It is such a reading of Foucauldian poststructuralism that has led some scholars to brand him a nihilist (Sarup 1989), a pessimist, a prophet of doom. Others are ambivalent if not outright skeptical about the utility of application of Foucauldian poststructuralism for educational research (Covaleski 1993: 6).

The picture, however, is not as bleak as it may seem. Poststructuralist theory has immense value in addressing issues and problems that are central to the educational process today. The archeological method, for instance, can be used to investigate how certain educational discourses (such as those of curricula and textbooks, science, social studies, etc.) came about in the first place. Similarly, the genealogical method can help us understand particular historical discourses by focusing on what can be said, who can say what is said, which "truths" are validated and legitimized, and what is excluded. As McDonough (1993: 2) argues, "Foucault's point is precisely that what we take as 'ordinary and valuable' is also often dangerous, and therefore always needs to be "critically questioned" (emphasis original). He goes on to argue that Foucault (following Nietzsche) advocates the use of the genealogical method that asks us to question how we came to believe what we believe (McDonough 1993). Cherryholmes (1987) makes a similar point while advocating the use of poststructural criticism for curriculum design. According to him, "If post-structural criticism teaches nothing else it is to be suspicious of argumentative, knowledge and policy, and policy claims based on appeals to precision, clarity and rigour" (Cherryholmes 1987: 309). To this it can be added that it also teaches one to be suspicious of universal, singular, and noncontingent claims to "truth." This is perhaps the single most important aspect of poststructuralism for educational research.

In order to examine the constitution of gendered subjectivities in the context of educational discourse in Pakistan, I use poststructuralist lens. I believe that poststructuralist theoretical perspective is more adequate to account for context specificity, interrogation of power relations, dynamics of exclusion and inclusion, unpacking of "regimes" of truth and reality, and specific techniques and strategies of historically situated educational discourses and their role in constituting subject and subjectivity. The use of poststructuralist theory requires that each time it is used, its political representations must be defined. It also follows that, since power relations are at the heart of such theoretical formulation, any application must also focus on the product of these relations.

Poststructuralist conceptualizations of power/knowledge and of discipline provide an analytical framework to examine and understand how the subject (the poststructuralist counterpart of Enlightenment "man") and subjectivities (contingent, mobile identities) are constructed within and by educational discourses, and where to look for vistas of resistance and agency.

In order to examine how gendered identities are formed in and through educational discourse in Pakistan, my main focus is on locating and identifying the constitution of subjectivities within the power relations inherent in such discourses. The relevance of poststructuralist theory for my inquiry is that it allows me to think of power as separate from and connected to the subjects, subject positions, and subjectivities just as the feminist appropriation of poststructuralist theory allows for gender to be separate from and connected to sex. Let me briefly engage with the poststructuralist notion of power/knowledge in relation to the instruments through which it is constituted and used by the operations of disciplinary control in Pakistan. Through this engagement, I show how curricula and textbooks should be regarded as important instruments of formation, recording, and dissemination of official truths.

## Power/Knowledge

One of the major contributions of poststructuralist theory especially in its Foucauldian hue is its rearticulation of the notion of power. Power in poststructuralist sense is a relation between various forces. It is not singular in terms of those who possess power (i.e., no single force possesses it). Nor is it singular in terms of direction, i.e., from top to bottom or vice versa. The exercise of power (for example, in the form of violence) does not constitute power: it is a *consequence* of power. Power in its modern form does not act as a constraining force that is corporeal in nature. In fact when "bodies" have certain legal rights that prevent exploitation and in situations where these bodies are rather free from direct forms of control, what needs to be explained are the methods through which power extracts time and labor from these modern bodies.

Power in this sense is made up of instruments for the formation and recording of knowledge (registers and archives); methods of observation, techniques of registration, procedures for investigation, and so forth. Once these instruments and techniques are developed and become entrenched, they can be appropriated and used by any institution of the society including the educational institutions, bureaucracies, and other administrative agencies.

Power has to be understood in relation to knowledge. Though power precedes knowledge, they imply each other. In other words, there can be no power (or power relations) without a simultaneous and correlative constitution of a field of knowledge. Similarly, there can be no knowledge that is not grounded in and constitutive of power and power relations (Foucault 1979).

What is crucial to understand here is that such an exercise of power does not suppress identities. In fact, according to the Foucauldian scheme of things, such an exercise of power constitutes subjectivities, e.g., "child," "madman," "deviant," and, in the case of this book, "women." Foucault admits that while the intention of the power could have been to produce regularity, the effect, more often than not, was quite the opposite, i.e., a multiplicity of identities (Foucault 1977).

Every society has its own regimes of truth that in turn allow which discourses are allowed and which are not. Currency and legitimacy of truth in a given society also depend on who owns and produces the truth. The ideological battle over the ownership and production of truth is an ongoing one. Different societies have different sites of truth production. In the Western context, for instance, it is the universities, armies, media, and writing that produce and own the truth.

In Pakistan, an important instrument of power is curricula and textbooks. These are used by operations of disciplinary power for formation and recording of knowledge as well as for its transmission. Controlled strictly and guarded jealously by the state, curricula and textbooks are one of the instruments that produce official knowledge. Official knowledges produce official truths. They work as instruments of normalization that continually try to manipulate people into "officially desirable" forms of thinking and behaving. A salient feature of the official truth partly produced in/through textbooks is the gendering of social, political, and sexual relations between subjects and between the subjects and the state. This feature is, however, also one of the least researched and examined aspects of power in Pakistan. I believe that it is imperative that we examine the ways, methods, practices, and techniques by which the official knowledge/ discourse (through curricula and textbooks in this case) normalizes populations. Similarly, it is also important to examine the ways in which the official knowledge deprivileges contending and competing forms of knowledge.

## The Genealogical Method

From within the Foucauldian poststructuralist tradition, the genealogical method is of particular relevance to the study of gendered subjectivities in/through the educational discourse in Pakistan. The genealogical method, with its emphasis on rejecting the primacy of origins, ideal significations, and original truth, helps me understand that knowledge does not simply uncover preexisting objects: it actually shapes and creates them. It also helps me understand that, since knowledge is not independent of the objects that it purports to study, it must be understood in its social and political context. The genealogical method comes to us from Nietzsche via Foucault where the former implores us to investigate the hidden and the secret history of a word in order to unearth the concealed context of the development of a concept.

Genealogy as an analytical approach is concerned more with the relationship between discourse and practice in history itself than with the practices of researchers and analysts in meaning making. The genealogist thus looks at traditional narratives of and about "historical developments" in order to discern the historical record, but from a position that is antihistorical (especially when seen from a traditionalist standpoint). As Dreyfus and Rabinow state, the "genealogist is a diagnostician who concentrates on the relations of power, knowledge, and the body in modern society" (1982: 105). Furthermore,

> the genealogist is to destroy the primacy of origins, of unchanging truths. He seeks to destroy the doctrines of development and progress. Having destroyed ideal significations and original truths, he looks to the play of wills. Subjection, domination, and combat are found everywhere he looks. Whenever he hears talk of meaning and value, of virtue and goodness, he looks for strategies of domination (Dreyfus and Rabinow 1982: 108–9).

The genealogist is not looking for any deep underlying truth or structure waiting to be discovered. Nor does she believe that there is an objective viewpoint from which discourses or society could be analyzed. Following Foucault, she believes that discourses became accepted as the dominant explanation of human experience that come to be shaped and controlled by the very discourses that purport to explain it. In Pakistan, educational discourse (among other discourses) does not simply inculcate in its subjects knowledge about themselves and their immediate and removed surroundings. It actively shapes and creates the subjects and their subjectivities. It is, therefore, important for me to carry out a genealogical analysis of Pakistan's educational discourse in its historical, social, and political context in order to trace the development of notions, signs, and concepts that are at the center of the educational discourse and that need to be investigated for their technologies and strategies of subject and subjectivity constitution. A genealogical analysis in this sense, for example, will be useful in analyzing how the narratives that seek to explain the Islamic and Pakistani self of the people actually constitute that very self in particular and peculiar ways that are in consonance with the power relations in the society. In Pakistan's case, a genealogical analysis is also instructive in that it leads us to understand the unfolding of different discourses and then (as in the case of the discourses of religion and politics) the symbiotic fusion of these discourses over time.

## Subject

Also central to poststructuralist thought is a reenvisioning of the concept of "man" or "individual." This reenvisioning is based on rejection of the Enlightenment notion of "man" as a metaphysical being, an independent, sacred, and singular category with inalienable rights and as one whose mind is a vista of true meaning and value. Calling it a "subject," Foucault tells us that "man" is constituted discursively and culturally through interaction with other beings and discourses and is both situated and symbolic at the same time. Each subject occupies various sites of meaning (differentially or simultaneously) that are culture based. Each of these sites (schools, family, workplace, social and/or marginalized groups) produces or induces varying identities, uses of language, configurations of self (and other), motivations, foci, and levels of agency. Thus, unlike the Enlightenment "man," the subject is constituted by and is part and parcel of the material world, its practices, mechanisms, and structures.

The poststructuralist subject derives her meaning and identity from the meanings and identities of the members of the identity groups in which she is located at given points of time. Within a given historical society, she also derives her meaning and value from the multiple common reservoirs of cultural symbols, relationships, and practices that she shares with members of cultural and subcultural groups.

It is important to understand that, unlike the sociological understanding of discourse as a means through which the subject expresses herself or accomplishes something, poststructuralist understanding of discourse is underlined by the assertion that the rules and criteria of discourses set up the conditions and spaces within which subjects are constituted or formed as teachers, students, managers, and so on.

The poststructuralist subject in contrast to the Enlightenment "man" is constituted across a number of power relations that are exerted over her/him and which s/he exerts over others (Foucault 1990). In other words, none of these power relations alone shall be singled out as the constituting one.

Subjection of the subject is rooted in particular historically located, disciplinary processes and knowledge that enable her to understand herself as an individual subject and that restrain her from thinking otherwise. These processes and concepts (or techniques) are what allow the subject to "tell the truth about itself" (Foucault 1990: 38). In this sense, the poststructuralist subject is not autonomous and sovereign. She is the product of the historical conditions that made different and various subjects possible in the first place.

Furthermore, the subject is a material effect of power and power relations when they install themselves. The material effects, one of which might be a particular subject, in turn act as channel for the operation of power itself. In turn the subject created by the power or the power relations herself becomes part of the mechanisms of power that had created her in the first place as the vehicle of power. The point is not to ignore the subject, but rather to examine the subjection and the process of subjection, i.e., the process of constitution of subjection. The subject in this sense has to be seen within and as a collection of techniques of power that run through the whole society and/or a particular social body. What is important is not those who have power but the actual field of power (Foucault 1980).

Poststructuralist subject is decentered. She, unlike her positivist and post-positivist counterparts, is not universal (and centered). Following in Foucault's footsteps, this decentering is based on the argument that external identities that are stipulated by the discourse are repressive and thus if one holds to them, no real positive change is possible. To move forward, it is essential to recognize the (historical and theoretical) contingency of the incumbent order of knowledge. For Foucault, histories must be thought and rethought so that an articulation that includes what has not been articulated and said shall come out. This is one of the major arguments for the decentering of the subject. As he writes in Power/knowledge:

The individual is not to be conceived as a sort of elementary nucleus, a primitive atom, a multiple and inert material on which power comes to fasten or against which it happens to strike, and in doing so subdues or crushes individuals. In fact, it is already one of the prime effects of power that certain bodies, certain gestures, certain discourses, certain desires, come to be identified and constituted as individuals. The individual, that is, is not the *vis-à-vis* of power; it is, I believe, one of its prime effects. (1980: 98)

Since it is the discourse, power, and power relations that constitute the subject, it is these that need to be the center of an inquiry rather than the subject. Doing otherwise might take our attention away from the constituting discourses. Consequently we might lose sight of what has been (and can be) said and what has not been (and cannot be) said.

Foucault, however, at no point attacked and undermined the idea of a subject that has knowledge of (and a relation to) itself. It was simply that such a version of the subject was not pertinent to studies of the historical emergence of modern forms of incarceration and the legal, social, and religious surveillance of the subject. However, I believe that the subject is indeed worthy of study when looking at the policing of sex or, for that matter, gender, especially if one is to start looking at the local level to determine how did some practices of power or the modern forms of institutions came about. For instance, in the case of Pakistan, regulation of sex during the 1980s and 1990s was more legal than ethical or social. Sexual activity became publicly accountable. In order to study this, it is imperative in my opinion to start with the subject who was at the receiving end of this legal and religious policing of sex during this time. In the same vein, in order to understand the working of the educational discourses, it is vital to take into account the experiences of the subject that has been constituted by these discourses in the first place.

Does the Foucauldian poststructuralist articulation of an unstable, decentered subject bode well for understanding the constitution of female subject? This and other such questions assume greater importance in the context of the claims made by feminists about the relative gains that various feminist movements have made over the years. Brown (2000: 73) presents a succinct summation of these questions: a) "Accustomed as we are to locating moral agency and political possibilities in a stable subject, how can we make sense of his talk here (re: decentering of the subject)? b) What subject understands the convergence of institutions resulting in the prohibition of research which would enable persons to better their institutions? c) Why would resisting the stable, unified subject be beneficial in the construction of political strategies? Those feminists who loosely follow Simone de Beauvoir argue that the "subject" has historically been a male and at "the margins of history is a female consciousness who must adopt certain male characteristics to be recognized as a subject in her own rights."

Foucault ignores this double history and thus his decentering of the subject has to be modified in order to accommodate the female subject. From a feminist perspective, it has been argued that Foucault's subjectivication is abnegation, a denial for men, and a suicide for women. Furthermore feminist scholars such as Harstock (1987: 160) have argued that poststructuralist theory, with its insistence on fragmented subject, came at a time just when women and other marginalized subjects were finding a common voice.

Foucault and those who follow him definitely need to be engaged on these and other such questions and criticisms. Does, for example, the notion of a decentered subject negate the gains made by and the "common voice" of the feminist movement? For me this requires a two-pronged engagement. First, listening to the voices at the margins of the feminist movement itself, we know that black feminists, women of color, Third World Feminists, and others have challenged the notion of a unified feminist voice. They point out to the appropriation of the voice by the white, middle-class feminists and argue that their experiences of oppression (and thus challenges to oppression) are very different from those of the mainstream feminist movement. It is in these voices that we hear arguments about the temporal nature of the feminist alliance. It is this temporal nature of the alliance that binds feminists together into a larger cause and at the same time allows them to retain their specific identities, to deal with their specific issues of marginality grounded in their specific experiences of oppression.

While it is true that Foucault does not offer an account of patriarchy, it is also true that he challenges the essentialist notion of patriarchy. What he shows is that while patriarchy did exist, for instance, in both Greece and Rome, one was very different from the other. While in the former women were treated as property, in the latter they were treated as a variety of objects for men. This asks us to investigate the form that patriarchy takes in its specificity in the contemporary era and in different societies. It is the patriarchal discourses and their historical specificity that can lead us to understand how the subjects of these discourses are constituted differently in different societies at different historical junctures.

Importantly, we need to look at the absent discourses by looking at the genealogy of the discourse, i.e., what has been recorded about a particular concept in the relevant historical time/context. This is also related to the Foucauldian notion of exteriority, which does not mean the objects outside of a subject, but rather outside of the prevailing discourse (or within which one is working) and even more importantly the exterior that becomes visible at the juncture where different (competing) discourses collide.

# Agency

There are two further questions that need to be asked of Monsieur Foucault in the context of a decentered subject. First, with its insistence on constitution of the subject and her subjectivity by discourse and its position on exteriority, does poststructuralism leave any room for resistance or agency and, thus, change? Second, with its emphasis on the fragmented nature of subject, what possibilities of struggle exist? From a poststructuralist perspective, while the disciplinary power multiplies its centers and localities, it also produces sites of resistance that were not there before. These centers of resistance, however, should not be understood as mere reactions to disciplinary power. Resistance is never in a position of exteriority to power. The resistance that acts as a counterpower coexists with power within the discourse in the context of power relations (Foucault 1979). Viewed from this position, resistance is an immediate reaction to or critique of the instance of power closest to it. The protagonists of struggle/ resistance (e.g., women vs. men; citizens vs. bureaucrats) do not hope to find solutions in a distant future. They rather look for them in the here-and-now.

Poststructuralist position with respect to the question of the possibilities of struggle by the fragmented subject is that just as the state devises strategies for harnessing the microforces of power into general strategies, so do the points of common resistance. These, however, cannot be attributed or attached to a single set of positions or objectives. In other words, while larger coalitions are possible, they cannot (or should not) assume the form of a metadiscourse of resistance, for that would be self-defeating (e.g., the Western feminist movement was not truly representative). This point has been of particular interest to poststructuralist feminists who have used it to respond to the feminist criticism of Foucault. The latter have raised the following question: If there cannot be a unified global feminist movement, then what is the panacea for the problems of women? Sawicki has used the Foucauldian position on the coming together of resistance to this effect. According to her, "depending on where one is and in what role (e.g., Mother, lover, teacher, anti-racist, anti-sexist) one's allegiances and interests will shift. There are no privileged or fundamental coalitions in history, but rather a series of unstable and shifting ones" (1988: 30).

Thus, for resistance to be effective, it should be directed at techniques of power rather than on a vague, broad, and general notion of power. Resistance, however, has to be reconceptualized. With the poststructuralist rejection of metadiscourses, it can be argued that the focus of resistance should shift from global to local. Resistance can no longer be conceived in terms of the fight against metaentities such as the state or capital. Resistance can be seen and should be conceived at local levels, presenting a "multiplicity of genealogical researches" (Foucault 1980: 83).

Foucault's arguments suggest a focus on praxis "within specific sectors, at the precise points where their own conditions of life or work situate them (housing, the hospital, the asylum, the laboratory, the university, family and sexual relations)" (1980: 126). In the context of my interest in understanding the constitution of gendered subjects in the educational discourse in Pakistan, the above articulations of subject, subjectivity, and agency guide me to look at the school, curricula, and textbooks. These are not only constitutive of the techniques and exercise of power but also the sites where I look for possibilities of resistance and agency.

This focus takes me beyond the traditional structural-functionalist and structuralist analyses of gender identity formation that do not allow any agency to the subjects constituted by the structures (discourse in my case). These analytical frameworks also obscure the normalizing role that curricula play. Poststructuralism, with its notion of dispersed power that is exercised and not possessed, thus helps me to expose the relationship between power/knowledge and gendered subjectivities that are normalized by and through textbooks. This relationship is by and large (if not completely) invisible both in the texts as well as in the analysis of education in Pakistan. In other words, poststructuralism helps me: a) to investigate the process through which gendered subjectivities are constituted by the discourse that is normally considered to be just talking about them; b) to explore what power do these subjects have in reference to the discourse that constitutes them; c) to examine under what conditions do these subjects resist the power; and d) to look for sites from which resistance is or can be mounted against the discourse.

*Space and Power*

Similarly, I find the poststructuralist articulation of space as power most useful for studying the constitution of gendered subjects and subjectivities in relation to the educational discourse in Pakistan. Poststructuralist conceptualization of this notion has several important implications for an analysis of gender and gendered subjectivities. First, space as power allows for effective, flexible, and detailed control of the body (the individual situated in a particular time and space) (Foucault 1979: 141–49). For example, an enclosure as a space where bodies are confined (e.g., home, schools, factory) allows for distribution of discipline as well as the "effective" management of bodies. Although Foucault had the example of Bentham's panopticon in mind, I argue that neither do the enclosures inevitably have physical demarcations nor do those confined need to be necessarily locked up. The enclosures can and do have symbolic confines, and those confined can have an ingrained fear/appreciation of the confines.

A good example of this is the symbolic confinement of women in *chadar and chardewari* (the veil and the four walls of the home) ever since the military regime of General Zia ul Haq in Pakistan. I call this the *panopticon of dress*. Under this use of space of power, women had relative freedom of movement in the social sphere so long as they were veiled and comparative freedom within the home. However, in both cases the space (veil and four walls) provided for "effective," flexible, and detailed control over them. The space is thus both real as well as ideal. It is hierarchical in both its physical and functional forms. As I show in my analysis of educational discourse in Pakistan, these spaces are not merely physically or architecturally constructed and constituted. The educational discourse, through the use of textbooks and curricula, discursively constructs spaces of power that in turn construct and constitute docile and marginal subjectivities and identities. Related to this is the notion of contexts of power. Poststructuralist edict refuses to explain away power solely in juridical, political or social terms. It does however stress an analysis of the concrete technologies of power. For example, looking at the correlation of punishment and crime, in the overall context of the shift from premodern to modern eras (shift from criminality of blood to a criminality of fraud) Foucault notes that punitive practices have been refined and fine tuned to have a more effective control over the bodies and subjectivities by means of surveillance of their day-to-day behavior, their identities, the apparently mundane gestures and body language, and so forth (1979: 77–78). These fine-tuned institutions become the contexts of power. Similarly, education as a technology of power (including classroom practices, curricula, textbooks, etc.) has also been refined over the years as to constitute fields of knowledge that are embedded in and constitutive of power and power relations. Education has also become an important context of power that needs to be investigated and interrogated.

Foucault suggests that "discipline produces practiced bodies; it increases the forces of the bodies (in economic terms of utility) and diminishes these same forces (in political terms of obedience)" (1979: 138). Such an understanding of the relationship between discipline and the constitution of useful and yet docile bodies provides for me, in this study, a specific analytical framework by allowing me to look for narratives of discipline in the educational discourse, particularly in curricula and textbooks.

### State and Power: Governmentality

In line with its conceptualization of power as scattered throughout the society, poststructuralism views and defines the state quite differently from the traditional explanations. The argument in a nutshell is that to view the state as an usurper of power and a solely punitive force is to walk the Marxist line of determinism and thus neglect the "technologies of power." It is further argued that "nothing will be changed if the mechanisms of power that function outside, below and alongside the State apparatuses, on a much more minute and everyday level, are not also changed" (Foucault 1980: 60).

Central to the Foucauldian conceptualization of the state and governmentality is the argument that it is misleading and erroneous to think that the state simply usurps the power mechanisms. These mechanisms have to be located and examined at the level of local, regional, dispersed panopticisms. In other words, we cannot just focus on the state if we are to fully comprehend the complexity of the mechanisms of power in their entirety. To do so will detract us from taking into consideration other mechanisms and effects of power that may not be linked directly to the state but that are as important to the state as its own constituting institutions that sustain its effectiveness (Foucault 1980).

In order to grasp the poststructuralist conception of the state, it is important to understand the Foucauldian notion of governmentality. The state, for Foucault, can be conceptualized as an "apparatus of social control which achieves its regulatory effects over everyday life through dispersed, multiple and often contradictory and competing discourses" (Burchell et al. 1991: 93). For him, the state initially was conceptualized as singular and external to the population. Thus, it was also conceptualized as fragile and perpetually under threat, since the external enemy wants constantly to dismantle it and the internal population has no a priori reason to accept its rule. As a corollary, the state remains locked in a constant effort to exercise power in order to reinforce, strengthen and protect itself (Burchell et al. 1991). To this was added the emphasis on territory and sovereignty (internal and external) in the seventeenth century. The conception of the state thus came to be dominated by the structures of sovereignty or the "reason of state." These structures, however, were circular, in that the end of sovereignty was sovereignty itself. It did not, per se, include the task of government. Following the juridico-legal and rights-based conceptualizations in the seventeenth and eighteenth centuries, a transition in the discourse of the state took place. It replaced a regime dominated by structures of sovereignty by one ruled by techniques of government. In this transition, however, sovereignty did not disappear. It continued to exist along with the exegesis of government.

Foucault terms this "governmentality," which is both external as well as internal to the state.[4] It is through governmentality that the continual definition and redefinition of what the state can or cannot do is achieved. The state can thus only be understood, in terms of its survival and its limits, on the basis of the general tactics of governmentality (Burchell et al. 1991). Government in this sense has to do less with governance of a territory or a population. It rather becomes a complex amalgam of people/subjects and things. The complexity of this amalgam lies in the way people/subjects are related to things and vice versa. As Foucault explains:

> Things with which in this sense government is to be concerned are in fact not men, but men in their relations, their links, their imbrication with those other things which are wealth, resources, means of subsistence, the territory with its specific qualities, climate irrigation, fertility, etc.; men in their relations to that other kind of things, customs, habits, ways of acting and thinking etc.; lastly men in their relations to that other kinds of things, accidents and misfortune such as famines, epidemics, death, etc. (Burchell et al. 1991: 95).[5]

There are, however, two things that need to be noted in the above conceptualization. First, Foucault does not refer to "men" in connection with gender, race, and class relations. Governance or governmentality not only regulates these relations in order to manage "men in relation to things" and in relation to others, but, in more ways than can be conceived of, it is in turn constituted and regulated by these complex relations. For example, the gendered and raced subjects that the state purports to regulate are also in particular and multiple relationships among themselves and with other subjects. These relations have their own exigencies and exert their own demands over the state and, in doing so, reconstitute the state. This is where the feminists "discipline" Foucault (Sawicki 1988; Bartky 1988) while appropriating him. They engender the subject, point to her race and other subjectivity, and thus invaluably contribute to the study of a multidimensional relationship between the subject and the state.

Second, the criteria of formation, transformation, threshold, and correlation of the discourses of the state and governmentality in the developing countries are starkly different from those in the West. Whereas it was the Christian pastoral model and the Westphalian military-diplomatic technique that ushered in the state and the discourse of governmentality in the West, in the world of the ex-colonies, the local discourses of state and government were disrupted by the imperial and colonial discourses of state and the government.

Third, in a Foucauldian conceptualization of the state and governmentality, men and things in relation to each other matter only in the internal context of the state. In the case of the former colonies, the criteria of formation were also influenced by how its men and things related to men and things outside the state. Thus one has to see the criteria of formation of the postcolonial states' discourse in relation to the metadiscourses of the Cold War, containment, the UN, international finance, international development, and so on.

Thus even if we do subscribe to Foucault's notion of governmentalization of the state, we might have to reinvestigate and redefine the notion in accordance with the criteria of formation, transformation, and correlation. Furthermore, this investigation also must be located in relation to the larger discourses with which the discourse of the state came into contact once the process of decolonization started. This is important because, more often that not, investigations regarding the state and other discourses in the postcolonial societies have been affected by the superimposition of larger discourses on the local ones (Abraham 1995).

The Foucauldian notion of governmentality is useful but insufficient for understanding the relationship between the state and the subject in ex-colonies. It is useful in that it differentiates between monarchical (physical) power and disciplinary power and brings forth the notion of governance in the contemporary period where the latter form of power accomplishes more than what the former used to do. His notion is also useful in that he rejects the view of the state as the repository of power, thus allowing for power to be seen as dispersed throughout the society and thus envisioning the possibility of resistance through the agency of the subjects. However, in the context of formerly colonized societies, this conceptualization falls short.

It falls short in accounting for the fact that institutions in these societies wield a peculiar and idiosyncratic mixture of monarchical and disciplinary power. The state in its various manifestations still has torture and death as the ultimate form of power over bodies at the same time as they use disciplinary power to normalize and create docile bodies. Use of physical and disciplinary power emanates from motivations linked to issues of domestic political legitimacy and demands of global economic regimes. The proportion of the physical and disciplinary power that comes into play depends on a number of factors that include but are not limited to the colonial interventions in the political, social, and cultural histories of different people; tensions between political legitimacy and international capital; and so on. Thus while the poststructuralist notion of governmentality points me in the general direction to investigate the relationship between discourses of state and education in constructing gendered subjectivities, I also use insights provided by poststructuralist feminists (below) and postcolonial theory (chapter 4) in my reconceptualization of state in Pakistan.

According to the former, instead of looking at the state as a monolithic entity that wields power and oppresses, it can and should be understood at various levels that are local (Bryson 1999; Watson 1990; Pringle and Watson 1992). It is at these levels that subversive activity should be directed (Bryson 1999). In this sense the state does not merely reflect and bolster gender inequalities; it constitutes them through its practices at the same time. As Pringle and Watson put it, "Gender practices become institutionalized in historically specific state forms" (1992: 64).

In the feminist debate on whether to retain or reject the state as an analytical concept, I take the position that it is more beneficial to look at the state in terms of an arena where contestation over subject, subject positioning, and subjectivity take place. There are then further subarenas where the state can be engaged, for instance, local government, state education, broadcasting, and so forth. The state in this sense becomes a site where meanings and identities are or can be contested, where engagement with institutions could take place. The state is thus important, albeit in a redefined manner: it is more useful to engage it rather than do away with it. As Bryson (1999: 100) argues, "From a post-modern perspective state agencies and institutions are also particularly important because of the ways in which they help construct and enforce the meaning of what it is to be a man or a woman in...society." Watson (1990) argues that the state, by responding to some demands and not to others, actually constructs these demands.

Recognition of the state in this sense is also important because only then can women hope and aim to gain access to and work from within the state, at whatever level, to make a difference. As the experience of femocrats[6] in Australia shows, this can mean success for the feminist cause, no matter how restricted or limited (Pringle and Watson 1992; Franzway 1986; Connell 1990).[7] It should, however, be noted that for femocrats or those working with the state, the basic objection was on the adjective "male" and not on the noun "state" (Allen 1990).

Poststructuralist feminists also argue that instead of looking at the state as a coherent entity that either influences a coherent set of interests that are formed outside it, or represents or embodies them, it is more useful to view it as a domain where such interests are articulated and defined (Pringle and Watson 1990). These interests are evoked and constructed through discourses. Furthermore, these interests are not unitary and do not represent unitary entities, e.g., men, women, and so on. "Interests are not merely reflected in the political sphere, they have to be continuously constructed and reproduced. It is through discursive strategies, that is, through creating a framework of meaning, that interests come to be constructed and represented in certain ways" (Pringle and Watson 1990: 229–30; also see Fraser and Nicholson 1990: 20; Nicholson 1990, Kenway 1995).

Discourses, in turn, are themselves constructed in this discursive space and construct the state at the same time. The discourses that construct the state assume a masculine subject rather than self-consciously defending or creating men's interests (Watson and Pringle 1990: 234).

Before moving on there are two questions that I would like to address. These are: a) Can males be feminist? and b) Can they be moral allies in the struggle against oppressions of various kinds?

It is important for me to answer these questions in the face of assertions from various hues of feminist thinking that males cannot really comprehend the realities of female suffering. By extension, it is asserted that it is extremely difficult if not impossible for men to investigate and/or interrogate the mechanisms of power that oppress subjects and also oppressive subjectivities. In order to respond to such assertions, I develop insider's advantage as a research strategy.

## Insider's Advantage and Male Moral Allies

The research strategy of Insider's Advantage takes its bearing from two philosophical notions: epistemic privilege (Mohanty 1995, 1997; Moya 1997) and moral ally (Blum 1999). *Epistemic privilege* refers to a "special advantage with respect to a possessing or acquiring knowledge about how fundamental aspects of... society (such as race, class, gender and sexuality) operate to sustain matrices of power" (Moya 1997: 136). Though it sounds essentializing, it is not. However, two cautions have to be heeded in this respect. First, as Moya (1997) has argued, while certain groups (usually oppressed) may have some epistemic privileges, social locations do not have epistemic or political meanings. Second, epistemic privilege in this respect is not analogous to the "standpoint theory," in that the advantage of epistemic privilege is only with respect to access and not necessarily in terms of proprietorship over episteme or knowledge. What epistemic privilege claims thus is "not any *a priori* link between social location... and knowledge, but a link that is historically variable and mediated through the interpretation of experience" (Moya 1997: 136; also see Mohanty 1995). In other words, social locations do not produce knowledge. They instead mediate between the social location of the subject and knowledge through the experience of the subject. For example, in research on postcolonial women, the social location of women lends them a voice that has to be heard and brought into the narrative. Similarly, the postcolonial researcher, because of her social location, also has access to knowledge that may not be available to others easily or at all.

Groups in privileged positions have made two common appropriations of epistemic privilege. The first is that of white males who, in the name of the Enlightenment reason, appropriated epistemic privilege, with the result that all others who did not have the "privilege" of being from this group (i.e., women, people of color, colonized natives, etc.) were disadvantaged and thus not capable of understanding. The second group to have appropriated this notion comprises women and oppressed groups. They have argued that in order to deconstruct and dismantle oppression, they have an advantage in terms of access to and possession of knowledge regarding the locations and operation of oppression in the context of class and gender.

Being a male, I do have the privilege of having knowledge (even if internalized, subconscious, latent, implicit, hidden, suppressed) of how power relations are constructed in a society like Pakistan and how they operate through various discourses. I also have the privilege of belonging to what is considered to be the elite club of males, and I use this advantage in conjunction with my post-structuralist feminist stance to unearth and examine how predominantly androcentric discourses of education, citizenship, and the state come about and also how these discourses construct gendered subjects and subjectivities that reproduce these discourses and are reproduced by them. The crucial question in this respect is whether I can use this advantage and privilege to conduct research that is nonrepressive (or, conversely, liberatory) and possibly even beneficial for women in Pakistan? Can I deconstruct my masculinity to an extent that I will be able to clearly understand the location of oppression in the narratives of the participants? My answer to this is, "I have at least tried." The use of "insider's advantage" gave me the following advantages.

Through this strategy, I was able to access information especially regarding identity construction and the interplay of gendered power relations from a male point of view. It helped me in accessing and interpreting the nuanced information because of my experience of being a male and also because of (as mentioned earlier) membership in the male club. This position also revealed, from a male perspective that is morally allied to feminism, how masculine subjectivities are constituted by the discourses and how in turn these subjectivities constitute discourses.

This is not to say that women are not capable of doing this (as has long been argued). The point is that in this regard, given the hierarchical social and cultural context of Pakistan, I have more chances and opportunities to obtain (particular) information from males than is likely to be the case with women researchers. Even more importantly, Insider's Advantage *is* a feminist research strategy. In my case, its importance also lies in the fact that males who are morally allied to the feminist cause have never used their maleness to the advantage of feminism. They have never used the advantage that their social location offers them to access knowledge that might be useful for the feminist cause.

To use insider's advantage from a feminist perspective, I draw upon Lawrence Blum's notion of the "white moral ally" (Blum 1999). The notion as it stands raises the questions: How can a white person relate to and help African Americans fight the oppressive racist system? If s/he ever can, then from what moral position can s/he do it? My question is: Can a male (albeit a feminist male) understand the reality of oppression of women and consequently try to help them? My response is, yes, as a moral ally I can.

# Chapter 3

# The Education System and Educational Policy Discourse in Pakistan

The history of educational policy making in Pakistan is one of laments, broken promises, and tall claims. Each successive government accuses the previous government of not being serious and committed to the educational cause of the nation. Each successive government promises to raise literacy levels to new (often unattainable) heights. Each regime promises to put in more money and allocate more resources to education. And each successive policy makes loud claims of harmonizing education with the principles of Islam. Yet, in real terms, education in Pakistan has remained largely underdeveloped both in quantitative as well as in qualitative terms. The questions that arise are: Why is the situation so bleak now and historically when it comes to educating the citizens? Are the governments, politicians, and bureaucrats not well-intentioned people? Do they intentionally make false promises? Are the educational planners actually not committed to the cause of raising literacy and educational levels and empowering the people of Pakistan?

Traditional explanations of the weakness, failure, and ineffectiveness of the educational system in Pakistan almost without exception agree that it is the ill-meaning politicians, patriarchal land-owning classes, and corrupt and self-serving bureaucrats who are the root cause of the dismal state of education in Pakistan (Qureshi 1975; Quddus 1979; Hoodbhoy 1998; Saigol 1995, 1993; Aziz 1993; Hasnain and Nayyar 1997). As a corollary of such consensus, some argue that preoccupation with political intrigues and political short-sightedness has resulted in the low priority accorded to education by successive regimes (both civilian and military) in Pakistan (Qureshi 1975; Hoodbhoy 1998). Some also argue that the intelligentsia in the new state is much to blame for not resisting the ruling elite's apathy toward education (Qureshi 1975).

Finally, those not too critical of the past or pessimistic about the future of education in Pakistan point to the political instability, enmity with India, defense problems, etc. for the lack of qualitative and quantitative increase in education and in literacy. As explained earlier (chapter 1) I do not wish to dispute the epistemic positions from which such explanations originate nor the conclusions they reach—I move the focus and raise different questions. For instance, following Ferguson (1990), I raise the question why the Pakistani politicians, bureaucrats and intelligentsia did not want the people of Pakistan to be educated. Why did the modernizing nation-state not want an educated workforce that could be a vehicle of development? Was it political shortsightedness, or are the traditional explanations of ill-meaning politicians and bureaucrats true? How can there be not a single well-meaning politician or bureaucrat in 63 years of Pakistan's history who could make a difference?

Following Ferguson (1990), my contention is that educational discourse in relation to other discourses—the Orientalist (i.e., colonizing and modernizing discourse) and the religiopolistic discourse in this case—constitute political as well as bureaucratic subjectivities that influence policies in postcolonial societies like Pakistan. I also shift the focus to looking at the state not essentially as a controlling possessor of power but in terms of governmentality whereby it uses education to harness biopower[1] and for purposes of surveillance and discipline.

Before going on to discuss subjectivity constitution in the educational discourse in Pakistan, I feel it is important to be cognizant of the contours of the educational system in Pakistan in terms of organization, decision making, policy discourse, and curriculum development. My main motivation in highlighting these features is to show that organizational and policy weakness is not the cause of the dismal state of the educational system in Pakistan as is often claimed by scholars and policy makers. To the contrary, it is a product of the educational discourse in relation with the discourses of state, nationalism, and so forth. While the gloomy state of the educational system in Pakistan affects all citizens, it relates even more strongly to women. To begin with, women are shortchanged by the system in terms of access to all levels of education. This inadequate access to education compounds the social and economic problems faced by women, especially those belonging to the lower social classes and minority groups.

Furthermore, those who do have access to education find themselves represented in the educational content as unequal to their male counterparts. Educational infrastructure, successive educational policies, and the curricula by and large contribute to according women an unequal citizenship status.

## The Educational System in Pakistan

Education is on the concurrent list[2] of subjects under the constitution of Pakistan. As such, the responsibility for education has been divided between the federal government and four provincial governments, namely, Punjab, Sindh, North West Frontier Province, and Balochistan.[3] Especially notable in this respect is the retention of curriculum and finance by the federal government. This is often seen by the provincial educational authorities as the desire of the federal government(s) to control the nature and direction of education.

### Literacy: The Numerical Dimension

At its birth in 1947, 85 percent of the Pakistani population was illiterate. In backward regions of the country, the literacy rate was even lower, with rural women virtually at zero literacy rate. Ever since, successive governments have declared the attainment of universal primary education as an important goal. Although considerable resources have been expended in creating new infrastructure and facilities in the last 50 years, the literacy rate in Pakistan remains low. Two-thirds of the population and over 80 percent of rural women are still illiterate. More than a quarter of children between ages of five and nine do not attend school. According to published government statistics, the current literacy rate in Pakistan stands at 56.2 percent (Government of Pakistan, Statistical Division, Online)[4]—an impressive 19 percent increase in the literacy levels in the last 15 years, from 37.2 percent in 1993, especially when plotted against the whopping population growth rate that has averaged around 2.33 percent a year in the corresponding period. The educational leadership of the country envisages literacy rate targets of 62 percent overall with 73 percent for males and 49 percent for females by the year 2015 under the EFA/MDG obligations (EFA 2009).

There is, however skepticism about the real value of these figures. It is a commonly known fact in Pakistan that official statistics are for the consumption of donor agencies and thus do not accurately represent the situation on the ground. This was stated by a number of people I interviewed at the Ministry of Education and the Ministry of Finance. The literacy and population growth rates are inflated and watered down respectively in accordance with the demands and pressures of donor agencies and the structural adjustment policies (SAP).

The reported literacy rates depend on what is used as the official definition of literacy at the time of their projection.[5] For example, during the Zia years (1977–1988), the official definition of a "literate person" was one who could sign his/her name. At other times a literate person has been defined as one who can read a vernacular newspaper. As one person whom I interviewed told me, "At times the officials responsible for projecting statistics start with a predetermined number or percentage given to them by the government and then keep on introducing dummy variables till such time that the desirable percentage is reached" (Field notes).

While the educational leadership has ambitious plans of raising the overall literacy level to 62 percent, gross primary enrollment from 83 percent to 100 percent, net primary enrollment from 66 percent to 76 percent, middle-school enrollment from 47.5 percent to 55 percent, secondary-school enrolment from 29.5 percent to 40 percent, and higher-education enrollment from 2.6 percent to 5 percent, the percentage of gross domestic product (GDP) that it has committed to education has declined. According to the UNESCO Institute of Statistics, based on the 2007 figures, the total resource commitment to the education sector is 2.8 percent of the GDP. Despite tall claims about removing gender disparity in education, the record in this respect shows that the gender gap in terms of educational institutions available for boys and girls as well as in terms of enrollment has increased instead of decreasing. The gender gap in terms of primary educational institutions in 1990–91 stood at 83,000 more schools for boys than girls. In 2001–02 this gap had increased to 109,000 schools. Similarly, in terms of primary enrollment, the gender gap from 2000 to 2007 stands at 78 percent. The gender gap in terms of secondary enrollment in the corresponding period stands at 77 percent (UNICEF, Info by Country [Pakistan]).

### The Infrastructure: Reproducing Inequality

Education in Pakistan is organized along primary (years 1–5), secondary (years 6–10), higher secondary (college; years 11–12) and degree (university; years 13–16+) stages. In terms of institutions, Pakistan has a multitiered system. On one level, these tiers are visible in terms of the public-private dichotomy[6] (or "partnership" as the development planners in Pakistan term it). Within each of these tiers are further divisions. Take, for example, the public-sector institutions. There are multiple tiers of public schools. The top tier are urban-based public schools that impart a better-quality education, are competitive to get in, and are prestigious. These include the Model schools (e.g., Islamabad College for Boys, Islamabad College for Girls, various Model Schools run by the federal government, the Divisional Public Schools, etc.). Also in the same league are the schools funded and run by the armed forces of Pakistan (e.g., various Pakistan Air Force, Navy, and Garrison schools), various preparatory schools-cum colleges that prepare cadets for the armed forces academies (e.g., Cadet College Hasan Abdal, Kohat, Burnhall, etc.).

The second tier is the urban-based government schools that are funded and managed by the provincial governments. The standard of these schools vary from one place to another. While some are better managed and impart a better standard of education, others are not as good. On the third tier are government schools based in the semirural and rural areas of Pakistan. These institutions often lack qualified staff and adequate infrastructure. Some are reported not to even have proper buildings. While most of the private schools in the top and second tiers are coeducational, public schools at almost all levels are segregated.

Private schools can also be broadly divided into three categories. In one category are the elite foreign schools such as the American and International schools in Karachi, Lahore, and Islamabad, and franchise schools such as the Choueifat School Lahore. In the second category are schools such as the Aitcheson College, Lahore; the Chand Academy; etc. These are very difficult to get into, and the fee structure is higher than the rest of the public schools. In the third category are private school systems such as the City School network and the Beacon House School system. These schools have maintained a higher standard of education, have fee structures that fall in between the elite and the public schools, are spread all over Pakistan, and in some cases follow their own curriculum. These schools also prepare students for the O- and A-level examinations conducted by the University of London. Finally, there are a large number of private schools that have mushroomed recently. These schools mainly cater to the lower-middle and lower classes and have low fees. The quality of education imparted by these institutions, while not very high, is still better than the government-run schools at the bottom of the public school system (Kardar 1998). Finally, later entrants in the schooling system are the NGO (nongovernmental organization) run and community-based public schools. Some of these schools, such as the Orangi Pilot Project (OPP) schools, have carved out a name for themselves. Others do not enjoy such a good reputation.

The educational infrastructure reproduces class inequalities at all levels. However, these inequalities are most evident and severe when they interact with gender inequality already existing in the society. While girls from the upper classes have relatively more access to education in quantitative and qualitative terms, those from the lower classes suffer on both counts. Girls in the rural areas of Pakistan are often the most affected.

## The Policy Discourse

In order to fully grasp the workings and mechanisms of the educational discourse in constituting subjects and subjectivities in Pakistan, it is important to understand the policy dimensions of the discourse. Constraints of scope and space do not permit a full and detailed discussion of the educational policy dynamics. I will, thus, briefly outline the main contours of the educational policy discourse.

Education policy discourse in Pakistan has been largely guided by the transposition of an educational vision that is grounded in the colonial and the Orientalist discourses of education on the one hand and by the global modernization and developmentalist discourses on the other. In the former sense, the emphasis and focus of the educational policy making was essentially focused on quantity rather than on the quality and relevance of education to the needs of the society. The underlying assumption was that once a critical mass of "literate" and "educated" population was reached, the society at large would benefit from an "improved" quality of life. In other words, it has been assumed that quality will follow quantity and that both quality and quantity will be in synchrony with the ethos of the society in the new state.

Broadly speaking, the colonial educational discourse is manifested in Lord Macaulay's famous (or infamous) Minute on Indian Education where he said:

> I would, at once stop the printing of Arabic and Sanskrit books; I would abolish Madrassa and the Sanskrit College at Calcutta....We must at present do our best to form a class who may be interpreters between us and the millions we govern; a class of persons Indian in blood and

colour, but English in taste, in opinions, in morals, and in intellect. To that class we may leave it to refine the vernacular dialects.

This thinking was conditioned by three factors: one, the exigencies of the global economic system of which the British colonial economy in India was a part; two, the demands of the colonial administrative system in India that required a class that could act as a cohort of the colonial administration; and three, the colonial cultural project.

Educational discourse in postindependence Pakistan drew heavily upon colonial educational discourse. In addition, as explained earlier, the political/nationalist leadership largely articulated the policies in light of their Orientalist understanding of the new nation and statehood where the main aim of the educational policy was to create a class of administrators (the civil bureaucracy) and a labor pool that could keep the economy of the new state in line with the demands of the peripheral capitalist system. At the same time, the religious discourse contested for and intervened in educational policy making and manifested itself in the liberal sprinkling of religious metaphors and references to the relevance of religion for the educational system of the new state (see, for example, Rahman 1953).

Education policy as framed by these discursive influences came to a head in the late 1950s in the form of the report of the Commission on National Education (Government of Pakistan 1959, hereafter GoP 1959). Saigol terms this report as the Magna Carta of education in Pakistan, "as it laid the fundamental structure which essentially remains intact, despite ideological shifts during subsequent periods" (1995: 118). The commission was comprised of representatives of big business, leading industrial houses, senior bureaucrats and officers of the armed forces in addition to consultants from leading Western universities such as Cambridge, Indiana, and Columbia, and think tanks such as the Ford Foundation and the Carnegie Institute.

The report laid down the basis for harnessing biopower by defining the citizen both as a worker and a patriot. It called for a curriculum reform aimed at developing "the basic skills in reading and writing and arithmetic, a liking for working with one's own hands and a high sense of patriotism" (GoP 1959: 115). At the same time the report also laid down the groundwork for discipline and surveillance. It specified the kind of workers the country needed, namely, the executive class, supervisory personnel, and skilled clerical workers (GoP 1959: 92). It also demarcated the spaces that the citizens were to occupy or be confined to. Educationally, unskilled workers were to be produced by elementary schooling, engineers and executives by professional colleges, and skilled technicians/clerks by part-time apprenticeship institutions (GoP 1959).

There are two things evident here. First, the educational discourse organized the system in terms of a mental-manual binary (between workers and the executive) and the hierarchical division of educational space where particular types of institutions were to produce particular types of educated subjects (e.g., elementary schools: unskilled workers, polytechnics: skilled workers, professional colleges: executives and supervisors). Second, it is noticeable that no mention of a *democratic* citizen or the institutional arrangements to produce such is evident in such arrangements. The emphasis was on producing *patriotic* citizens. Patriotic citizenry is easier to control and discipline than the democratic citizenry. Could it be as Saigol (1995) argues, that this emphasis arose because a military government was running the country? I argue that this was not the sole reason. Educational discourse in the new state had produced such a line of thinking even earlier. Both Jinnah and Fazlur Rahman (who presided over the All Pakistan Educational Conference convened in November 1947) stressed the importance of technical and vocational education along with spiritual and civic education (see GoP 1947; also see Rahman 1953).[7] The emphasis on technical and vocational education was aimed at producing depoliticized workers for the economy, whereas that of spiritual and civic education was to inculcate nationalistic and patriotic values, as was evident in the curricula that were developed following these guidelines.

It is not too difficult to gauge the impact of such policies on women. The mental-manual binary itself was exclusionary. It excluded women from the manual workforce that the policies envisaged to develop. For example, most of the polytechnic institutions were exclusively for males. On the other hand, the limited opportunities for female students in institutions that purported to develop the "mental" workforce effectively meant that females were also excluded to quite an extent from this category as well. The inculcation of nationalistic and patriotic values also worked differently for different genders. For females it meant inculcation of values such as sacrifice and a supportive role in contrast to the more active and public patriotism of their male counterparts. The system in this sense worked to strengthen the public-private binary in gender relations in the society.

Drawing upon colonial educational discourse, the educational discourse in postindependence Pakistan aimed at depoliticizing the student-citizen. The 1959 report stipulated that "students should not participate in politics, or serve the interests of groups outside the academic community in activities inimical to the orderly conduct of the institution and its academic program" (GoP 1959: 58). In other words, students were to remain confined to the intellectual space constituted for them and were not expected to have an opinion on political matters of the country. The discourse thus aimed at creating passive subjects. The emphasis was on maintaining order rather than creating faculties of critical thinking and citizenship.[8] Other than the brief interlude of 1972–1977, this depoliticization of the student population has carried on until today. Students unions are banned in the country by the order of the Supreme Court since the time of General Zia ul Haq's martial law (1977–1985).

The influence of the religiopoly in constituting docile subjects who could be easily ordered in order to harness their biopower was omnipresent but increased over time. For example, *Islamyat* (literally, "study of religion," but in practice, the study of selective history and teachings of Islam) was made compulsory up to class VII by the 1959 education policy. The 1972 policy made it compulsory up to class X and the post-1977 policies first extended it to classes XI and XII and then made it compulsory for the professional colleges as well. Let me take a moment to explain the link between teaching of Islamyat and the constitution of docility and orderliness. The subtle (and later not-so-subtle) subtext of *Islamyat* textbooks emphasized subservience not only to God but also at all levels from family head to the ruler of the day. It also emphasized the virtues of war, patriotism, and nationalism and created the binary between the good and the bad citizen based on these criteria. It can safely be said that nationalism and religion became synonymous.[9]

Post–Ayub Khan era education policies retained the aims of the previous policies largely articulated by the economic discourse of modernization and peripheral capitalism and the religious discourse. The 1970 policy, for instance, mimicked the 1959 report by stating:

> Without harnessing the vast human resources available to Pakistan, the task of sustaining and accelerating economic development remains unfulfilled. In this regard the basic objectives are, on the one hand, to broaden rapidly the base of education with a view to attaining the ideal of universally literate and productive society and, on the other ensure a continuous supply of highly trained and creative leadership. (GoP 1970)

The 1972 education policy defined the ideal Pakistani citizen as "dynamic, creative, capable of facing the truth: an individual able to comprehend fully the nature of technical and social change and having deep concern for the importance of society (GoP 1972: l). Educational policies during the nonplan period like the previous policies also focused heavily on discipline as one of the main objectives of education. There was, for instance, a greater (and different) emphasis on uniforms. Uniforms served as the mechanism of constituting particular identities through inclusion and exclusion. Zulfiqar Ali Bhutto's education policy, for example, mandated that all boys were to wear the national dress (*shalwar kamiz*) to school. The declaration of *shalwar kamiz* as the *awami* (national) dress laid down one more criterion for nationalist identity. However, only men and boys were supposed to wear *shalwar kamiz* to work and school respectively (though some urban upper-middle-class women also did). Women and girls were excluded from this visual representation of nationalism and thus the citizenship criterion.[10]

During 1980s and 1990s, the framing of educational policy drew heavily on the global political and economic discourses marked by Reaganism and Thatcherism, the religiopolistic discourse at home, and the Gulf Syndrome (also popularly known as the *Dubai Chalo*: "Let's go to Dubai" syndrome). Influenced by these discourses, the educational discourse in Pakistan constituted a subject that immigrated to Gulf countries, got rich, and now contested for political, economic, and social space with the existing privileged subjects at home. The subjectivity of this Dubai returned subject was conservative, nationalistic, and religious. These subjects, in their contestations with the existing subjects in Pakistani society, in turn affected the nationalist and religious discourses that were instrumental in their symbiotic fusion.

In this period we see religiopoly making a significant intervention in the educational policy realm. As mentioned earlier, *Islamyat* was made compulsory up to the professional college level. In addition *mohalla* (neighborhood) schools and *madrassahs* (religious schools) also became a part of the mainstream educational discourse. The 1979 education policy, for instance, sought to establish five thousand *madrassahs* for boys and five thousand *mohalla* schools for girls.[11] Owing to the increased influence of religiopoly, segregation was the name of the game even at this level. What is important is the implication of these schools and their segregated nature. Since *madrassahs* trained boys in religious studies and since, as noted earlier, religion and nationalism had come to be synonymous, by implication the boys received a better education than girls did. Also by implication the boys thus were to graduate as better citizens. The influence of religiopoly has continued even after the death of General Zia and the ushering in of democracy in Pakistan, contrary to what most scholars believe.

A recent education policy makes special provision for *madrassah* schools. According to the policy document *"Deeni Madaris* [plural of madrassah] are independent institutions. They have organized themselves into 5 *Wafaqs/Tanzimes/Rabitaes*. There are about 7000 *Madaris* in Pakistan with the total enrolment of about one million students including 78000 females. The Government of Pakistan is providing incentives to integrate religious education with the formal education and to bring these *Madaris* into mainstream" (Government of Pakistan, Ministry of Education, Online).[12]

Education policy discourse also drew upon religiopoly to create a vision of a homogeneous society where the homogeneity was solely based on religion, which had come to be synonymous with nationalism/patriotism. Thus, all differences (ethnic, cultural, gender, etc.) were obfuscated. While minorities were excluded outright from the "nation," other groups such as women, ethnic groups, and cultural and linguistic identities were also by and large excluded from the notion of nation and citizenship. A good example of this is evident in the 1983 education policy. Other than institutionalizing *madrassah* and *mohalla* schools (without any consideration to such things as curricula), the policy mandated that university degrees at the graduate and postgraduate levels should not be awarded to the candidates unless they have imparted the basic Quranic *Qaida* (primer) *Yasarnal Quran* (GoP 1983).

What it effectively meant was that no minority member could get a graduate degree because he/she was in no position to teach the Quranic primer and that the definition of "literate" would now include persons who could only visually read Quranic Arabic (even without really understanding it). This reminds me of my French professor who used to open the French course with the remark that by the fourth week of the course the students will be able to read French fluently as we read the Quran, i.e., without understanding it.

What is also interesting to note is the language and symbolism that made its way into the educational policy discourse. The 1984–1988 policy documents, for instance, appellate literacy projects as *Iqra* project, *Sipah e Idrees* (Army of *Idrees*), and *Razakar* Muslim project. In all three instances, the symbols used are religious. In one instance, the reference is to view teachers (and students) of a literacy projects as an army. What is also important to note is that the degrees issued by the *deeni Madaris* on the basis of arbitrarily set curricula were made equal to the degrees awarded by national educational institutions. This practice continues till to date.[13]

In this period, the education policy also mandated that all *Hafiz e Quran* (those who have memorized the Quran by heart) will be awarded 20 extra points when competing for admission to professional colleges and universities. Those with paramilitary training (known as National Cadet Corps Training [NCC]) were also to receive 20 points. Thus, military and religious training enabled one to have a better chance of entering professional colleges and universities. I, as a faculty member at the departments of international relations and defense and strategic studies, repeatedly came across a number of candidates who had bogus *Hafiz e Quran* certificates. The candidates possessing *Hafiz e Quran* certificates often failed to recite verses from memory when asked to do so.

The 1979 education policy stated:

> The highest priority would be given to the revision of the curricula with a view to reorganizing the entire content around Islamic thought and giving education an ideological orientation so that Islamic ideology permeates the thinking of the younger generation and helps them with the necessary conviction and ability to refashion society according to Islamic tenets. (GoP 1979)

The effects of the discursive articulation of education policy in Pakistan also permeated other realms. For example, those attempting the civil service exams needed to rote learn verses of the Quran and Arabic prayers, without which they could not pass the interview stage of the competition. Similarly, in interviews for government jobs, candidates were asked to recite Quranic verses and prayers and were judged just for their memory and not their understanding abilities. During Zia ul Haq's reign, the knowledge of Quran and *Sunnah* (the way of the Prophet Mohammad) was also made "necessary for appointments of teachers. It was suggested that all textbooks would be reviewed by a committee of religious scholars who would be required to remove any content repugnant to Islam" (Saigol 1995: 182).[14] The discourse also constricted physical and intellectual spaces through emphasis and decree on use of *dupatta, chadar*, and *hijab* in educational institutions as I explain in chapter 7.

It must be noted, however, that the "Islamization" of education was not meant to create an Islamic education, social, or cultural ethos. Nor did it mean that it was necessarily a harbinger or companion of an Islamic economy. It was mainly aimed at constructing docile subjects. It was also aimed at normalizing certain categories and naturalizing certain subjectivities by emphasizing the masculine.

Although the educational discourse in Pakistan in its various policy manifestations advocated equality of sexes and gender, in reality the binaries that were so inherent in the discourse (e.g., mental-manual, objective-subjective, patriotic-unpatriotic, feminine-masculine) permeated all levels of education from segregated educational institutions to suitability of subjects for boys and girls. It also further strengthened the inequity and disparity in gender relations. From the time of the National Education Conference in 1959 to the Report of the Commission on Education in 1959 to the latest education policies, the educational discourse has constituted gendered subjects and subjectivities.

Where and whenever the policy discourse talks about improving educational opportunities for women, it essentially is influenced by the demand of creating a workforce that is necessary for harnessing the biopower of the female subjects in relation to that of men. This is especially the case under the Structural Adjustment programs. This is why education, instead of improving the plight of women, has burdened them with the double responsibility of working at home (with no help from men) and outside at the same time, while men are expected to work only in the public sphere. All educational policies emphasize women's role as mothers necessary to take care of the male workforce. Similarly, the vocational training for women specified and recommended in these policies was also geared towards this end.

## Curriculum Development in Pakistan: Constructing Gendered Nationalist Subjectivities

According to the constitution of Pakistan, education is a provincial responsibility while it is on the concurrent list. However, in reality, as mentioned earlier, being on a list of concurrent subjects implies that the federation and the federating units (the provinces) share the responsibilities as well as the expenses. From these responsibilities the federal government has kept finance and curriculum development under its jurisdiction. The Curriculum Wing (CW) of the federal Ministry of Education is the body responsible for curriculum development and for providing the provincial ministries with education guidelines according to which textbooks, teaching guides, and other materials are to be produced.

The basic rationale given for this arrangement is based on the argument of standardization of texts all across the country. This quest for standardization of curricula is in turn grounded in the necessity of nation building. In other words, the quest for standardization is basically a quest for homogenizing different cultures, ethnicities, languages, and religions in accordance with the notion of a single nation on the site of textbooks. This notion of a single homogeneous nation in itself is grounded in the Orientalist notion of the nation-state that emerged in Europe after the treaty of Westphalia in 1648.

The CW thus produces standardized curricula that obscures any or all differences and in doing so also lays down the criteria and conditions for citizenship. As I show below, the criteria constructed in and through the text produced in accordance with the curricula are gendered and thus constitute gendered subjects and subjectivities. The only provision for the accommodation of difference comes in the shape of supplements to Social Studies textbooks for each district.[15] These supplements cover the local history, geography, demographics, and important personalities of the area. Even in these supplementary texts, these topics are portrayed in a nationalist and gendered way. The supplement, for instance, on the district of Attock does not talk about the Buddhist period of its history and only lists important persons in terms of those who were instrumental in the Pakistan movement. These personalities are always male and nationalistic.

Furthermore, as stated on the Ministry of Education website, one of the main aims of the CW is related to "directing any person or agency to delete, improve, or withdraw any portion or whole of the curriculum, textbooks and reference material prescribed for any class being repugnant to Islamic Teaching and Ideology of Pakistan" (MoE, Online). (Note the use of capital letters in Islamic Ideology.) In these functions of the CW, we see the predominance of the religiopolistic discourse. It is also evident that it purports to constitute subjects that are nationalistic and patriotic through a homogenizing notion of "nation" and Islam. Though no systematic data is available in the form of textbooks and curriculum advice from the years prior to 1990s, from whatever textbooks are available it is evident that, historically, curriculum and textbook development has followed the trajectory of the modernizing influence of nationalist and religious discourses to a point where these discourses fused into a symbiotic relationship. In other words, until such time that the nationalist and modernizing discourses were dominant, religionization of curricula and the resulting texts were not as pronounced. However, since the symbiotic fusion of these discourses, the religious and the nationalist content has also fused in a way that the two have virtually become synonymous.

I want to conclude by answering the fundamental questions about educational policy discourse and especially curriculum development raised in the beginning of the discussion in this chapter. The questions asked were: Are the educational policy makers in Pakistan (bureaucrats, politicians) ill-meaning, self-serving, and blind to the interest and well-being of the people of Pakistan? Are the curriculum developers and the textbook writers really oblivious and blind to the gendered subjects and identities they create though the school curricula and textbooks? Do the policy makers, curriculum developers, and textbook writers really want a gendered and obscurantist version of Islam to frame subject and subjectivity constitution in Pakistan? My response to all three questions is negative. I do not believe that all policy makers, curriculum developers, and textbooks writers have narrow personal, political, or ideological agendas that motivate them to play havoc with the educational policy making in the country.

I believe that these people, who are competent in their respective fields, are themselves products of different discourses present and working in the society. In fact they are the subjects constituted by the same discourses that through the policy they make, the curriculum they develop, and the textbooks they write constitute other subjects and subjectivities in the society. However, in arguing so I do not in any sense want to imply that they as subjects of the discourse are helpless (because this would amount to falling into the same structural trap that I want to escape) or that there is absolutely no intentionality involved.

My argument is that the subjectivity of these policy makers is also constituted by the same educational, economic, and political discourses, locally and globally, that constitute other subjects. They like other subjects have agency and the capacity to resist and even change the discourse. That their agency has only been able to bring about small changes in the educational structure, system, curricula, or textbooks (evident in relative increase in enrollment of female students, a miniscule increase in the female presence in textbooks, etc.) should not be taken to be signs of total helplessness. In the final analysis, as Ferguson (1990) has argued (in the case of international development planning), it is the discourse (in this case educational) that constitutes the subjectivity of educational planners and thus determines the policy they formulate. As I have shown in the above discussion, educational policy in Pakistan has followed by and large the emergence, disappearance, trajectories, and relative power of the discourses of the state, religion, and nationalism.

# Chapter 4

# Women and the State in Pakistan: A History of the Present

From 2004 to 2009, a total of 1,635 women were murdered in what were claimed to be honor killings: 143 were killed by fathers, 448 by brothers, 334 by husbands, 61 by in laws, 177 by relatives, and 57 by sons. Out of the 898 FIRs (First Investigation Reports) registered, only 354 persons were held (Human Rights Commission of Pakistan [HRCP] website).

In the same period, 3,669 cases of rape were reported; 1,898 were gang raped. Only 481 accused were ever held (HRCP website).

On average, at least two women were burned every day in domestic violence incidents. Approximately 70 to 90 percent of Pakistani women experience spousal abuse. A woman is raped every two hours in Pakistan, and in Punjab, a woman is raped every six hours and gang-raped every four days (HRCP website).

As many as 88 percent of female prisoners are serving time for violating the 1977 *zina* ordinance, which makes fornication a crime and adultery a state offence (National Commission on the Status of Women in Pakistan 1995).

The above-mentioned statistics are not peculiar to Pakistan. Almost all South Asian and many other countries exhibit similar oppression of women. The reason for citing these statistics is not to paint a victim image of Pakistani women; rather, the purpose is to show the extent of violence and its gendered nature, in the face of which Pakistani women have used their agency to their own advantage. In this chapter, I historicize women in relation to the state in Pakistan. My main purpose is not to provide a historical context for these notions but to trace the contours of the discourses of gender relations and the state through the Foucauldian strategy of archaeology, i.e., how these discourses came about. I also briefly trace a genealogy of these discourses to uncover the processes of inclusion and exclusion, fixation of meaning (moments), creation of a body of floating signs (i.e., elements whose meanings have not been fixed), and how these elements mount resistance to the fixation of meaning of the signs.

I have two motivations in undertaking an archaeology of the discourses of gender and the state in Pakistan. First, the theoretical framework (outlined in chapter 2) informs me that traditional historical analysis either displaces or ignores those events, subjects, and practices that do not fit within a model of continual historical development. Archaeology in its Foucauldian articulation critically and discursively examines such historiography. As a method, it aims to understand the meaning-making practices of those who create discourses; it also seeks to uncover those rules of discourse that allow such meaning making to take place in the first instance. Second, I believe that undertaking a concurrent archaeology of the discourse of gender in relation to that of the state shifts the focus away from personalities (of leaders) and points to underlying rules that brought about the discourses in the first place. Doing so also has the added advantage of explaining the timing of emergence (and disappearance) of certain discursive practices that often remain unexplained in the literature on gender and state in Pakistan, as I argue below. The contours of gender and the state traced in this chapter frame the discussion of the constitution of gendered subjectivities in and through the educational discourse in Pakistan in chapters to follow.

In order to historicize women and the state in Pakistan, I draw upon two bodies of literature. The first of these is what I term *statist literature*. Studies in this body of literature deal with a variety of subjects and dynamics of the state-society relationship in Pakistan.[1] This literature is rich and varied. It emanates from a number of different epistemic positions ranging from Marxist to historical-structuralist, revisionist historiography, civil society, and institutionalist.[2]

The second body of literature that I draw upon in order to historicize gender and state in Pakistan is comprised of studies on the relationship between the state and women in Pakistan. This scholarship has been produced by feminist scholars and activists located both within and outside Pakistan.[3] As I mentioned in the first chapter, my intention is not to show that the conclusions reached by scholars who have produced these bodies of literature are wrong or faulty. My main intention is to reformulate the questions that have been asked by the scholars working in these areas. I thus draw upon the work of these scholars to show that recasting the questions from a different epistemic position leads to a better understanding of how gendered subjectivities are constituted in Pakistan.

## Conceptualizing the State in Pakistan

In order to historicize the relationship between women and the state in Pakistan, it is important to take into account various conceptualizations of the state there. Here I consider three influential ways in which the state in Pakistan has been conceptualized. These are the historical-structuralist, Marxist, and postfoundational explanations.

### Historical-Structuralist Explanation

Central to the historical-structuralist explanation of the state in Pakistan is the claim that the authority structure of the postindependence state in Pakistan is largely an inheritance from the British colonial state in India and thus an ideal entry and focal point to analyze politics and the relationship between the state and the society in Pakistan (Waseem 1994).

Also central to this conceptualization is the notion that the state is largely, if not completely, autonomous and that there is continuity in the location of power from colonial times to the present (Waseem 1994: 3). In the case of Pakistan, it is argued, the state is more than mere law or a law-making apparatus in that it comprises both a "formal institutional apparatus and a vast network of patron-client relations performing an informal control function" (Waseem 1994: 446). Thus, the Pakistani state has been able to accommodate mass dissent and open its doors to various subelite and underprivileged groups while remaining a vehicle of the elite classes (Waseem 1994: 449).

Inherent in such a conceptualization is the Weberian polar-dichotomous notion of modernity, where the path to modernity runs from the premodern to the modern in continuous succession (see Evans et al. 1985; Skocpol 1985). The colonial experience of the colonized societies thus becomes one of the many stages through which the society (and the state) passes or has to pass en route to becoming modern. Also inherent in such conceptualization is a model of state that resembles the modern European state in that it possesses power and the institutional means of control and ideological and political integration of various classes and groups in the society. It is thus the state that regulates the power relations (including gender(ed) relations) in the society.

## Marxist Explanation

The second major conceptualization of the state in Pakistan is grounded in the Marxist tradition. Hamza Alavi, in his ground-breaking article "The State in Post-colonial Societies: Pakistan and Bangladesh" (1972), argued that in the case of post-colonial[4] societies, the Marxian analytical category of one dominant class is not valid. Instead he proposed that in post-colonial societies, there is more than one economically dominant class, such as the metropolitan bour-geoisie, indigenous bourgeoisie, landed bourgeoisie, and so on (Alavi 1972). He further argued that none of these economically dominant classes is *the* dominant class. In the case of Pakistan, for instance, the role of the military-bureaucracy (not essentially one of the eco-nomically dominant classes) cannot be overlooked. Alavi locates the post-colonial societies in the structural matrix of peripheral capital-ism, where any particular combination of the dominant classes (with or without the help of the military-bureaucracy oligarchy) uses state power to extract and accumulate wealth in order to serve peripheral capitalism (and not capitalists in particular) (Alavi 1972, 1990). The post-colonial state in this sense, according to Alavi (1990), might not even be the main extractor of wealth. It is most certainly not autono-mous. The regulation of power relations in the society is not solely a function of the state; it is rather a function of one or more dominant classes working to the benefit of the peripheral capitalism.

Seen from this perspective, it is the complex configuration of the landlords, the indigenous bourgeoisie, the *mandi* (wholesale mar-ket) merchants, the *mullahs* (religious leaders) and the *pirs* (spiritual leaders), the salariat, and the military-bureaucracy oligarchy that constitutes the "dominant" class in Pakistan. This configuration of dominant classes works for peripheral capitalism (in the name of the state) and in doing so defines power relations in Pakistani society (Alavi 1972, 1990). It is both important and interesting to note that Alavi's conceptualization of the post-colonial state in Pakistan— which on its appearance triggered a vigorous debate on the subject— does not directly deal with the issue of gender. Owing to his Marxist orientation, Alavi assumes that gender is subsumed into classes and that once the situation improves with respect to class relations, the situation for women will improve automatically. However, feminist literature on women in communist societies shows that this did not happen.

Both historical-structuralist and Marxist accounts of the state in general and the Pakistani state in particular share a common notion of the state, i.e., an idealized form fashioned after the European nation-state carved out in Westphalia in the seventeenth century. As Ashish Nandy notes, "All political arrangements and all state systems are . . . judged by the extent to which they serve the needs of—or conform to—the idea of the nation state. Even the various modes of defiance against the state are usually informed by this standardized concept of it" (2003: 5). Problems that plague these societies are seen from this vantage point, and the solutions are dependent on the veracity of state or its withering away, depending upon the epistemic position from which the analysis is done. However, few post-colonial states or societies have evolved along the Westphalian model (Nandy 2003).

### Postfoundational Conceptualization of the State

The third epistemic position from which the state in Pakistan can be conceptualized is what may be termed the postfoundational position. However, very few analyses from this position have been undertaken in the case of Pakistan. Scholars such as Khattak (1994) analyze the state from this perspective, but her analysis in particular often slips back into the foundational trap.[5] I will discuss two aspects of the postfoundational articulation of the colonial project in some detail as they inform the historicization of women and the state in Pakistan. The first is the colonial knowledge and discourses that produced shifts in social relations and political control in the colonies. The second is the question of how these discourses, as appropriated by the postindependence nationalist leadership and intelligentsia, continue to understand the state and the subjects in Orientalist terms.

Colonial intervention altered social identities and organization in the colonies in such a way that new social categories, identities, and dichotomies were constructed where they did not exist prior to colonization. For example, in the case of India, as Dirks shows so persuasively, "caste was refigured as a 'religious system', organizing society in a context where politics and religion had never before been distinct domains of social action" (1994: 8, emphasis original).

The same can be argued in the case of Islam in India, which was reconfigured by the colonists as an organizing system by bifurcating family from civil law so that the "community" was given power over the family while the colonial state retained power over the community. Dirks summarizes this phenomenon succinctly:

> The success of colonial discourse was that, through the census, land-holding, the law, inter alia some Indians were given powerful status in new formulations and assumptions about caste, versions that came increasingly to resemble the depoliticized conditions of colonial rule. (1994: 9)

The discourse of the modern nation-state appropriated by the postindependence nationalist elite drew upon the colonial discourse in this respect and thus addressed problems and issues of development essentially along the lines of the secular vs. religious dichotomy. Together the Orientalist and modernization discourses have served not only as the sources of knowledge about the postcolonial states and societies but (more importantly) also as a means of ordering the basic categories and assumptions that we employ in our analysis. The Orient or the postcolonial world that we live in or seek to understand is, as Said (1978) argued, a product or effect of the collaboration of knowledge and power mediated by colonial intervention. Orientalism[6] in this sense is more than the colonial imagination and economic interests: it refers to a sophisticated body of knowledge that constitutes subjects and subjectivities even when the colonial period ends.[7]

Because of their inability to get out of the Orientalist perspective, nationalist scholars and leaders assume that the postcolonial societies are just as unitary and homogeneous as societies in Europe once were and that modernization will follow independence automatically. These assumptions gloss over diversity and difference in the name of unity and leave out those individuals and groups who do not fit into either the modernity or the nationalist project (Prakash 1992; Chatterjee 1986). Both colonial and nationalist historiography leave out events and discourses for the sake of the coherence and unity of the historical text or narrative. More often than not, women, minorities, and subaltern groups, and especially their experiences, events that affect or involve them, and/or their histories do not become a part of the historical narratives of either the colonial or the postcolonial state (Dirks 1994).

Various accounts that trace the history of the state in Pakistan (or of the society, or of groups within the society) operate from the Orientalist position outlined above. However, here I would make a distinction between feminist and mainstream scholarship. Feminist scholarship in Pakistan, while essentially couched in the Orientalist categories of thought,[8] particularly focuses on the histories, narratives, and experiences of women and to a lesser degree on those of minorities.[9] Mainstream scholarship, on the other hand, is Orientalized both in terms of its epistemic positions and its gender and subaltern blindness.

Following Prakash (1992), I historicize women and the state in Pakistan through a critical reading of traditional historiography that has relied to date on certain foundational categories such as nation, state, class, and so forth. I focus on relationships and processes that have constituted contingent and unstable identities both at the individual and group levels. I try to bring in events that have been left out previously. I also reread gender-blind texts to uncover the gendered nature of these texts.

In studying the relationship between women and the postindependence state in Pakistan—following Foucault—I move the focus from the patron-client relationship between the two to the mechanisms of discipline and surveillance that the new state employs over subjects. I also add to this the focus on representational forms through which the discourse of the new state constitutes gendered subjectivities (Guha and Spivak 1988). Thus, in a way I move the focus from the relationship between state and women to the relationship between subjects constituted by the discourse of the state.

## The State and Women in Pakistan

The emergence of Pakistan as an independent entity is articulated by scholars in a variety of ways.[10] Notwithstanding the epistemic positions from which these articulations have been made, what is common to perhaps all of them is the assumption that Pakistan or the state in Pakistan is a continuation of the British colonial state both in terms of time and space. Also inherent in these articulations is the assumption that colonialism was a historical phase in the progression of history from Mogul India to the independent "modern" nation state of Pakistan. A resultant assumption is that the colonial legacies have been transmitted (in a sense of linear continual progression) from the colonial state to the successor state of Pakistan.

While scholarship based on these assumptions does provide useful insights and entry points into an understanding of the way the relationship between the state and social groups in Pakistan is understood, it glosses over the fact that the colonial intervention in India changed the categories of thought through which Indians (and later Pakistanis) came to understand the relationship between state and society (among other relationships). For example, the linear view of history does not account for the changes that the British colonial project brought about in the lives of a number of subaltern groups. Similarly it also excludes the narratives, events, and histories of a number of groups from the historical narrative of the independence movement and from that of Pakistan. In short, what can be termed the official history of the independence movement and of Pakistan has become a history of certain elite groups (usually male, usually middle and upper class, and always Muslim) who are argued to have brought independence to the Muslims of India and who have since then been the custodians of the nation's security and development.

Since the scope of this chapter does not mandate a detailed examination of the colonial intervention in India or how it changed the way Indians and Pakistanis write their own histories, I focus on examining the impact of British colonialism on Muslim women in India and in Pakistan after 1947 and specifically on two aspects of this. First, I look at how the colonial intervention and discourses privileged some groups (Muslim men in this case) by formulating assumptions about the place and status of other individuals and groups (Muslim women in this case). Second, I examine how the colonial discourse on women interacted (in terms of contestation and interdiscursivity) with the nationalist discourse after 1947 to fix the meaning of woman and womanhood in Pakistan.

## Women and the Colonial State

British colonialism in India is often credited with improving the condition of the "subjugated" Indian woman. The most often-cited example is that of the colonial banning of the practice of *sati* (widow immolation) in India.[11] The conclusion drawn is that, by banning a "barbaric" act, the British colonizers empowered women in India.[12] However, what remained until recently underresearched is the colonial motive for doing so, which had nothing to do with altruistic (and humanist!) motives. Postcolonial and subaltern studies scholars point out that, by shifting the focus from altruism to the cultural project of colonization, the motivations behind the apparently "civil" and altruistic acts of the colonizers become clearer.

Furthermore, as Spivak (1993) shows, what is conspicuously missing is the subjectivity of the Indian woman immolated through *sati*. Spivak suggests that there is no space in the colonial discourse from where the female subaltern (the *sati* women in India) can speak. She also shows that while the voice of the male native elite can be found (recovered) in (from) the colonial "text," that of the "brown, Indian" woman cannot.[13]

Although the impact of colonization is visible and significant in almost all aspects of the lives of the colonized people in India, two areas are especially important with respect to the articulation of gender. First, the colonial economic discourse brought about changes in the land administration system that altered the economic as well as the social and political reality of the Indian populace in general and had a deep impact on the status of Indian women, both Hindu and Muslim.[14]

The new land administration system also altered the balance of gender relations. As Mumtaz and Shaheed argue, "The new colonial land administration actually reinforced retrogressive feudal and tribal structures in the rural areas, and by giving the feudal landlords and tribal heads absolute ownership of the land, increased their own power and that of the elite ... [thus] ... strengthening the subordination of women in those areas" (1987: 36).[15]

Similarly, the changes brought about by the colonial administration in the legal realm also had a far-reaching impact on gender relations in Indian society. Let me cite some specific examples of how these changes affected the Muslim women in undivided India. While the legal system in India was Anglicized in almost all its realms, such as criminal law, land tenure, evidence, and so forth, the realm of family law governing interpersonal affairs—including those between men and women—was left up to the community (Chatterjee 1993; Mumtaz and Shaheed 1987).[16] In consideration of the fact that the interpretation of Islamic law had always been a male realm, the colonial discourse privileged men over women in matters such as marriage, divorce, maintenance, guardianship and custody, depositors of property, and so on (Mumtaz and Shaheed 1987: 37).

With respect to the Muslims of India, this created a complex and contradictory situation. On the one hand, the colonial administration stripped the community leadership from the dispute settlement by monopolizing the dispensation of justice (Waseem 1994), while on the other, it gave the men of the community the right to adjudicate the interpersonal affairs of the community that in turn entrusted the task of interpretation of Islam to the *ulema* (religious scholars) (Metcalf 1990). Furthermore, the colonial state, by implementation of the Hindu customary law regarding inheritance of property, deprived

Muslim women of the right to inherit property also. They could only administer property on behalf of minor sons (Mumtaz and Shaheed 1987). Muslim women were thus rendered nothing more than repositories of Muslim culture. Their seclusion, marked by *purdah* (veiling), became an issue of identity for the Muslim community.[17]

Before moving on to the discussion of women and the nationalist state in Pakistan, let me touch upon an aspect of the colonial discourse that has had a far-reaching impact with respect to the articulation of the status of women in Pakistan. By fixing the meanings and boundaries of social categories (such as communities, e.g., Hindus, Muslims, etc.) in religious terms, the colonial discourse on the one hand legally and discursively formalized Hindus and Muslims as communal rivals while on the other constructed binary division based on secular/religious definitions. The formalization of Hindus and Muslims as communal rivals, according to one scholar, contributed to Muslim demands for separation articulated in, for instance, the two-nation theory (Brass 1979). The secular/religious binary on the other hand framed the contestation between the nationalist leadership and the religious elements all through the independence movement and ever since, so that today the state and the major political parties are still pitted against the religious right. Thus, the state has often been seen as a natural partner of women against the religious and obscurantist forces. Feminist analysis of the state, however, shows that there is nothing natural about this so-called alliance.

### Women and the Nationalist State

Until the appearance of feminist scholarship in Pakistan (mainly since the late 1970s), women appeared in the history of the independence movement merely as footnotes. Their discussion comes either in the context of a few educated, elite women serving as appendages to male leaders such as M. A. Jinnah, Suharawardy, and Sir Shahnawaz or through passing references to isolated acts of gallantry in the face of the colonial authorities.[18]

The nationalist discourse prior to and after the creation of Pakistan was articulated mostly in terms of the Western discourse of nationalism. Although the stated basis for the demand and creation of the new state was the provision of a separate homeland for the Muslims of India, the meaning of "nation" was closer to the Western sense than to the local context of nationalism. In this respect, the nationalist discourse was not a counterdiscourse to colonialism, but rather a reaction that emerged from within the colonial discourse and was thus embedded in it.

The nationalist leadership, while successful in dismantling colonial rule, could not dismantle the epistemic confines of Orientalism. Thus, like the colonial state, the nationalist state also glossed over racial, ethnic, linguistic, and class differences in the name of unity and homogeneity. It took upon itself the mantle of the modernizing state but found itself contesting with social differences and realities that did not quite fit into the development discourse of modernity. Issues related to women were thus either to be dealt within the framework of nationalism (e.g., honor of women, abducted women, etc.) or within the modernization framework, i.e., as an issue of productivity (1961 family laws, etc.). To put it another way, the Pakistani woman was either defined through the Orientalist discourse (a sum of colonial and modernization discourses) or through the religious discourse where the meaning of being a woman was fixed by the *ulema*. It was thus a contest between the two discourses in the field of discursivity that fixed the meaning of woman.

An important factor in this respect was the resistance and agency of the women of Pakistan from within these discourses, which from time to time allowed them not only to forestall the discursive fixation of meaning but also to change these meanings. It was, for instance, the agency of the women of Pakistan that made religion a political issue right after independence. The issue at hand in 1948 was that of granting property rights to women—rights that had previously been denied to them by the colonial government—in the form of the West Punjab Muslim *Shariat* Application Act. The landlord-dominated Punjab Assembly was reluctant to pass the act as this would have given women recognition as independent entities in a patriarchal social order (Jalal 1990).

For the first time in the newly independent country, a few hundred *burqa*-clad (veiled) women gathered and protested in front of the Assembly Hall against the delays. It was perhaps the first of the few incidents in Pakistan where women used their subjectivity created by the religious discourse to their advantage. At the same time it was the first of many times that the women of Pakistan demonstrated their agency.[19]

Jalal's explanation is that the newly independent state and the nationalist leadership did not support the cause of women as they could not afford to alienate the dominant classes on the question of gender (1990: 87). A similar analysis has been offered by Rousse (1998), who argues that the state only conceded to the demands of women or acted on women's issues when it was either politically or economically expedient. Such explanations are sound if the focus of inquiry is the autonomous state and its self-maximizing state managers.

However, once the state is conceptualized in relation to the men and women it governs, such explanations fall short. An alternative explanation of the relationship between the state and women in Pakistan can be found in the discourses that constitute them. The modernizing state in Pakistan was located squarely within the discursive confines of the modernization discourse. At the same time, the subjectivity of the women of Pakistan was being constituted (and contested) by the religious, nationalist, and modernizing discourses. The subjectivity thus constituted was multiple and contingent. That is why we notice different subjectivities emerging at different points in the history of Pakistan. This also explains the nature and timing of the way the state has related to these multiple subjectivities. The analysts have often understood this as political expediency.

A good example of this contestation can be found in the framing and promulgation of the Family Law Ordinance of 1961. Two things are interesting to note in this respect: first, it was not only the *Ulema* who opposed women's efforts to be included in the citizenship discourse; and second, it is interesting to note the demands that the women of Pakistan were making in the period immediately following independence (roughly 1947–1955). The Charter of Women's Rights included demands that ranged from reserved seats in the legislature for women, to equality of status and opportunity, equal pay for equal work, and the guarantee of rights for women under the Islamic Personal Law of *Shariat* (Mumtaz and Shaheed 1987). The range of demands shows that women of Pakistan at this point were less concerned with religious or so-called secular discourses. They asserted their agency from within both discourses to demand what they thought would best improve their condition and status in the wider society. In doing so while the religious and the modernizing discourses constituted female subjectivity, women in turn played an important part in shaping these discourses. This is an important point to note because almost without exception, credit for drafting this women-friendly ordinance is accorded to the military government of General Ayub Khan, the first of the four martial laws in Pakistan (Waseem 1994: 159; Mumtaz and Shaheed 1987: 57 are especially conspicuous in this respect).

The Family Law Ordinance aimed at discouraging polygamy, regulating divorce procedures and registration of marriages, raising the minimum age of marriage from 14 to 16 for girls and from 18 to 21 for boys, regulating allowances after divorce, and so on. Though the ordinance was never really implemented, it had a major impact in two ways: first, it gave women a legal framework from within which to mount resistance to the patriarchy: and second, it shifted the mantle of regulating personal laws and the lives of citizens (especially women) from the *ulema* to the state. A third important but hitherto unnoticed effect of the ordinance was that, through records (for instance of marriages), the state now had the apparatus of surveillance that enabled it to extend its gaze to the family and personal space. In the next chapter, I discuss some of the discourses that made possible the surveillance of the female subjects.

# Chapter 5

## Subject Positioning and Subjectivity Constitution in Pakistan

### Women and the Constitutional Discourse in Pakistan

Pakistan did not get its first constitution until 1956. Since then, two more constitutions have been promulgated, one in 1962 and the other in 1973. The last mentioned has been hailed as the constitution on which there was a complete national consensus. For most of its life, however, the 1973 constitution has been suspended by respective military governments. I will not go into details of the constitutional issues here due to constraints of scope and space, but will instead concentrate on how constitutional discourse in Pakistan affected women.[1] In each of the articles of the constitution that aim at safeguarding women's rights, provisions to this effect have been explicitly stated. For instance, articles 25 and 34 of the 1973 constitution stipulate that:

  i. All citizens are equal before the law and are entitled to equal protection before the law.
 ii. There shall be no discrimination on the basis of sex alone.
iii. Nothing in this article shall prevent the State from making any special provision for the protection of women and children.

Furthermore, article 34 provides that "steps shall be taken to ensure the full participation of women in all spheres of national life." Article 27 stipulates that there will be no discrimination on the basis of race, religion, caste, or sex for appointment in the public services of

Pakistan. Furthermore, article 35 stipulates that the state shall protect marriage, the family, the mother, and the child. In addition to these provisions, the constitution gives women the right to vote and run for office (including the posts of president and prime minister). The 1973 constitution also reserved special seats in the legislature for women for a period of 20 years—a provision renewable by the nation's lawmakers.[2] In 1972, the government of Z. A. Bhutto had already opened all civil service positions to women. They could for the first time enter the district management group (the elite DMG), foreign service, and other government services. However, these constitutional provisions have not been able to provide the level of security to women that was intended. One of the reasons cited for the ineffectiveness of these constitutional safeguards for women is that the 1973 constitution was made ineffective by laws instituted by the military regime of General Zia ul Haq through ordinances. Another reason often put forth is that politicians have never been serious about implementing the constitutional provisions. I analyze these arguments in the following section.

## Religiopoly

I introduce the term *religiopoly* to mean a symbiotic merger of religious and militaro-nationalistic discourses where each discourse retains its originary criteria of formation but where these discourses together form the dominant discourse that constitutes subjects and subjectivities, positions subjects, spells out disciplinary mechanisms, and so forth. The symbiotic nature of the merger means that the discourses retain the capacity to compete and contend even within the symbiotic relationship and to separate at some point in the future. I introduce this notion to explain and show that the relationship between militaro-nationalistic and religious discourses has taken a new shape, that of a symbiotic interdiscursivity.

In poststructuralist terms, the intertextuality of these discourses reaches a point that I term *symbiotic interdiscursivity*: a state where two or more discourses, while remaining separate and distinct (and even in contention), draw from a common pool of meaning. It is important to understand that the discourses do not merge or fuse to form a new discourse. They retain their shapes, strategies, and mechanisms of meaning fixation while entering into a symbiotic relationship. Nor is this symbiotic relationship permanent: it is rather dependent upon the changes within each of the discourses. These changes take place in the discursive third space within these discourses.

This conception of the third space is different from the field of interdiscursivity in the sense that it is located within the discourses rather than outside them as in the case of field of discursivity. The third space can be understood as the space *between* the meanings of signs fixed within the discourses that are in a symbiotic relationship. This is how it works: discourses in a symbiotic relationship constitute subjects and subjectivities that are in particular relationship with each other. Meanings of signs (e.g., woman, man) are fixed discursively in each of the discourses. The space between the meanings of signs is the space in which meanings are contested (while the discourses are in a symbiotic relationship), which in turn results in changes within the discourses. These changes may be manifested in a rupture of the symbiosis or in its strengthening. This is also the space from where resistance is mounted from within the discourses. As I argue next, by the late 1970s, the religious, political, and militaro-nationalist discourses in Pakistan had merged symbiotically, and this merger set the discursive realm within which meaning fixation and subject positioning took place.

Just as Ayub Khan and Zulfiqar Ali Bhutto are credited with enhancing the status of the women of Pakistan (due largely to their liberal views regarding women), General Zia ul Haq is singled out as the person whose policies were most detrimental to the women of Pakistan. Zia ul Haq rode to power on the back of a popular mobilization of people whose demand was to implement *Nizam e Mustafa* (literally "the Prophet Muhammad's system"). Although the coalition that mobilized the people against the popularly elected government of Zulfiqar Ali Bhutto in 1977 was comprised of religious as well as secular parties, not all of whom were serious about the demand for Islamization, Zia made it the centerpiece of his rule.

It is important to understand the motivation of a large segment of Pakistani society in demanding an Islamic system. There is almost a consensus among scholars that the real reasons behind the mass mobilization against the government of Zulfiqar Ali Bhutto were socio-economic and not purely religious (Waseem 1994; Jalal 1990a, 1995; Alavi 1990; Noman 1990). The popular mobilization (led by the Pakistan National Alliance, a nine-party alliance of religio-political and secularist parties) is attributed to causes such as middle-class disaffection, the alienation of industrialists because of the nationalization of industries, and the disaffection of the *mandi* merchants.[3]

What seems to have been left unexplored is the fact that many people were insecure not only because of economic and political reasons but because of the severe and violent intimidation by the regime, especially during the last years of its tenure. While both men and women were targeted by the senior Bhutto's regime, women bore the brunt when the political opponents of the regime were faced threats to the security of the female members of their families (Jalal 1990). What has also not been taken into account in the analysis of this phenomenon is the practice of abducting young college girls at the behest of Pakistan Peoples Party (PPP) ministers, which became rampant toward the end of the PPP's reign. Thus, while on the one hand women became more visible, on the other hand their personal security was threatened. The first was fodder for the religious elements while the latter was a worrying point for the citizenry. This explains the willingness with which the people took to the streets in their protest against Zulfiqar Ali Bhutto's regime.[4]

This was the point in time when contestation over the fixation of meaning of what it meant to be a woman in Pakistani society came to a head. Islam and Islamic ideology became an important nodal point around which new meanings were to be fixed. I am not arguing that Islam was not a point earlier or that it was not important. My argument is that religion as a nodal point at this juncture is different from the role it played during the independence movement in the 1940s. During the independence movement, it was the nodal point of nationalism around which the meanings of different signs were fixed, and religion, while instrumental, was not the nodal point itself. At that time (around 1977 onwards) both nationalism and ideology merged symbiotically to form a nodal point around which the meanings of different signs were to be set by the discourse. This is also seen as the point in time when the alliance between the military and the *ulema* also crystallized. It is in this discursive context that Zia's Islamization fixed the meaning of the Pakistani woman as half that of a man, or in other words, as half a citizen.

The question that arises at this point regards the timing of the merging of the nodes of nationalism and ideology on the one hand and the (un)holy alliance between the gun and the pulpit on the other. Why was it possible after the 1977 martial law and not at the time of the two earlier martial laws in 1958 and 1969 or the subsequent coup d'etat in 1999? I argue that the answer lies in recasting what Rousse (1998) describes as the alliance between military and *ulema*, i.e., a kind of fusion between the politico-religious discourse and the discourse of the state. These two discourses, which have been drawing upon while at the same time contending with each other since the time of the independence movement, merged or fused at around this time in the history of Pakistan to create what I call religiopoly.

In the case of Pakistan, religiopoly represents the symbiotic inter-discursivity between the politico-religious discourse and that of the modernizing/nationalist state. By late 1970, this symbiosis was in an advanced stage. This explains the timing of the so-called Islamization under the dictatorial regime of general Zia ul Haq. More than an alliance between the military and the *ulema*, it was the symbiosis between the discourses that produced militarized and religio-political subjectivities. This is perhaps as good an example as one can find of what Foucault terms "power/knowledge."

Another question that can be answered on the basis of the framework of symbiotic discursivity between discourses or religiopoly is why the state in Pakistan delegated power to men, even if it was just power over women. As Foucault reminds us, power needs some resistance in order to be acknowledged as power. In the case of Pakistan, the delegation of power to men over women served a dual purpose. First, it acted as a modicum of power that helped the power/knowledge of the religiopoly to be recognized as such. Second, it diluted the sense of disenfranchisement that the people of Pakistan might have felt during the martial law years. And yet, within the third space there emerged the agency of a certain segment of Pakistani women in the shape of the Women's Action Forum (WAF). I discuss this in a separate section on Pakistani women's agency below.

In the field of discursivity, other discourses, such as the legal discourse or those of media and economics, drew upon the dominant discourse, i.e., religiopoly, as well as upon each other and constituted subjects. Gender and power relations between genders were one result of the subjectivities constituted through this process.

## Women and the Legal Discourse in Pakistan

Prior to General Zia's version of the Islamic legal system, the legal discourse in Pakistan for the most part drew upon the British (colonial) legal discourse. With respect to the legal status of women, the law provided them in most cases with a status equal to that of men. Their rights and responsibilities were also congruent to those accorded to men. Those areas where they did not enjoy equal rights, such as in inheritance (discussed above), were also incidentally the ones in which the colonial legal discourse denied women equal rights by invoking or not invoking religious laws of the respective communities. The legal discourse in Pakistan also drew upon global economic discourses including those of modernization and development to accord certain rights to citizens.

The cornerstones of Zia's Islamization in the legal realm were the *hudood* ordinances[5] and the *qanun e shahdat* (law of evidence).[6] These laws purported to repeal the older laws concerning theft and armed robbery, rape and adultery, and alcohol consumption. Two new laws—those of *qazf* and punishment by whipping—were added (Jahangir and Jillani 2003). The stated motivation of Zia ul Haq behind the promulgation and implementation of the religious laws was the need for reform of the legal system and for bringing it into conformity with Islam. However, as is evident from the selective promulgation of religious laws, reform was not really the intent. For example, in the economic realm, no serious effort was made to promulgate religious laws. The only area in which religious law was allowed to impact the economic sphere was in the question of interest paid to investors by banks. Here, two things were apparent. First, the introduction of so-called interest-free banking was nothing more than the renaming of interest as profit-loss-sharing (PLS).[7] Second, as far as the state's fiscal affairs were concerned, payment of interest on loans (mainly to international lending agencies and states) was permitted. Similarly, different agencies of the state, such as the House Building Finance Corporation (HBFC), continued to charge interest on housing loans and mortgages from its clients. Even in the case of laws such as *hudood* and *qanun e shahdat*, more emphasis was placed on punishment (discipline) than on reform.

Subject positioning in the legal discourse can be surmised from the fact that while there have been virtually no cases of the amputation of hands (a punishment prescribed for theft) or that of feet (a punishment prescribed for armed robbery)—crimes usually committed by men—despite the fact that both types of crime were rampant in Pakistan during this time, court cases where women have been punished (flogged/imprisoned) for offences of fornication or extramarital sex ran into the hundreds (Jahangir and Jillani 2003). Similarly, women also became the focal point of the *qanun e shahdat* ordinance. Though the Islamic law of evidence is more than just the stipulation that the evidence of two women equals one man (and even this is disputed by scholars of Islamic jurisprudence), this became the focal point both in its implementation as well as the impact it had on society in general. I will come back to this point, but first let me discuss how the *hudood* laws as a part of the legal discourse shaped the contours of the discourse itself and what impact they had on the subject positioning and subjectivity constitution of women in society at large.

It is a well-documented fact that the main purpose of the *Hudood* ordinances was not societal reform. Some excellent statistics and analyses show that, in its implementation, the laws targeted the subaltern population, especially women, members of the lower classes, and minorities (Jahangir and Jillani 2003; Mumtaz and Shaheed 1987; Jalal 1990, 1990a, 1991; Rousse 1994, 1998; Saigol 1995; Waseem 1994; Jamal 2002).[8] I draw upon these analyses to focus on how *hudood* laws, as a part of the legal discourse, created gendered subjectivities. In other words, my focus here is the meanings that were fixed by this part of the legal discourse for groups such as women, men, and members of minorities. My argument is that, by looking into the ways in which the discourse created gendered subjects and the relations between these subjects, we see more clearly how women have come to be understood and how they understood themselves. We can also see how the state used these legal provisions to keep an eye on and discipline subjects, whether men or women.

Among the *hudood* laws that became a part of the legal system in Pakistan, perhaps the most infamous have been the *zina* laws. *Zina* literally means "fornication," and the implication is that such an act is consensual. The *hudood* laws with respect to *zina* first of all remove the distinction between consensual sex, adultery, and rape. Prior to the promulgation of these laws, the legal system in Pakistan defined adultery and rape as criminal offenses while there was no penal sanction for consensual sex between unmarried adults. Under the *hudood* laws, all three forms of sexual activity were collated under *zina*. Under the pre-*hudood* criminal system, rape was a crime punishable for men alone,[9] the rape of one's wife was an offense, the consent of a child under fourteen years of age was immaterial, and children under seven years of age could not be punished for any offense. Punishment for rape was transportation for life or imprisonment extending to ten years including a fine, punishment for the rape of a minor wife (under twelve years of age) was the same as that for rape, punishment for marital rape (over twelve years of age) carried a maximum of two years, and punishment for adultery was either imprisonment for five years or a fine or both. There was no punishment for sex between consenting adults (extracted from Jahangir and Jillani 2003: 86–87).

In contrast to this, under the *hudood* laws, "women can be charged for the offense of rape. Rape on a wife is no offense. Children regardless of age can be convicted for *Zina* or rape. Consent of a child can be put forward as a defense by the accused. Where consent is established, the offense converts from rape to *Zina*" (Jahangir and Jillani 2003: 85). As compared to the maximum punishment of twenty-five years' imprisonment for rape under the pre-*hudood* criminal system, the new law provided a maximum sentence of ten years of rigorous imprisonment, thirty lashes, and a fine. Furthermore, minors can be imprisoned for five years and are liable to a maximum of thirty lashes should they be convicted for any sexual crime. The new law makes adultery and consensual sex between adults nonbailable offenses. Complainants cannot withdraw charges once a case has been registered. Finally, and perhaps most importantly, charges of *zina* can be made by any person, whether aggrieved or not (extracted from Jahangir and Jillani 2003: 85–86).

I argue that promulgation of the *hudood* laws, including the *zina* laws, was not merely the whim of a general. Nor was it just his quest for legitimacy or simply the result of his collusion with right-wing political parties like Jamaat e Islami. Had that been the case, the reaction to the laws would have been more pronounced than it was. I argue that religiopoly had become dominant to an extent where it had made such practices seem normal. Just as Hitler could not have done what he did without a society that had come to believe in the normality of Nazism, Zia could not have promulgated and implemented the laws in a society that had not already been normalized by religiopoly.[10]

What has been the impact of *zina* laws on Pakistani society in general and on gender relations in particular? Let me start with some implications of the new law. First, under the *zina* law, anyone can report and lay charges for a case of *zina* whether aggrieved or not. This can be interpreted in two ways: first, that any member of the society can now report an act that s/he thinks is an offense against the moral and religious fabric of the society; and second, that members of the society at large have become instruments of surveillance for the state. The application of the said law in Pakistan shows that the second and not the first has been the primary outcome (see cases reported by Jahangir and Jillani 2003: 89–130). Details of the *zina* cases registered show that husbands have registered cases of *zina* against their wives and former wives, that political opponents have filed *zina* cases against political rivals, land and property disputes have often resulted in the filing of *zina* cases, and that in a number of instances, resistance by women to sexual advances has resulted in the filing of *zina* cases against them by the perpetrators.

The second important point is that under the new law, women have to prove that the rape had been committed. Women face a double jeopardy in this respect. Consider the following scenario that I have drawn from a number of cases. A woman registers a case of rape. The case goes to court. The law of evidence in the case of *zina* offenses requires four pious Muslim males to testify that a rape actually took place and that they had witnessed it. The onus of producing such witnesses is on the complainant. Now here's the double jeopardy: in cases where such witnesses cannot be produced, evidence of rape cannot be established. Once a rape cannot be proved, the woman can be (and invariably is) charged with fornication since she has admitted that a sexual act did indeed take place. It is not hard to see the connotations of this for women in Pakistan. What it effectively means is that women are highly unlikely to have recourse to justice in the face of sexual exploitation and crimes.

Considering these two implications of the changes in the legal discourse in Pakistan, a picture emerges of the gendered subjects constituted by the discourse. By partially devolving the surveillance function of the state upon society in general, which in a patriarchal society like Pakistan effectively means men, and by removing the recourse to justice for women, the discourse created power relations in which the balance clearly tilted towards the male segment. Indeed under the *hudood* laws, even attempted *zina* is crime. Attempted *zina* is different from attempted rape, which was a crime even under pre-*hudood* laws. As to what could be considered preparation for *zina*,[11] this was left up to the courts to decide on a case–by-case basis. Furthermore, the religious courts kept pressuring the government to legislate upon preparation for *zina* as a crime (Jahangir and Jillani 2003: 115). As to how "preparation for *zina*" differs from "attempted *zina*," this becomes clear in the light of recommendations by the Federal *Shariat* Court (FSC). According to Jahangir and Jillani, "The FSC strongly recommended that under the system of Islamization a woman sitting in dubious circumstances with strangers (males) should be made a penal offense in Pakistan" (2003: 115–16). In other words, a woman sitting with men could be considered "preparation for *zina*." What is implied is that, so long as the woman is sitting with the men with their (men's) consent, it is all right. However, the moment consent is not there, she can be accused of "preparing for *zina*" and thus liable to be prosecuted.

Fortunately recommendations such as these for making "preparation for zina" a crime never became law. However, these and other such recommendations from time to time find their way into the media discourse, which also draws upon both religiopoly and (among others) the legal discourse. The social implications of the *hudood* laws have thus become a part of the process of meaning fixation especially for women. In a psychosocial sense, the female subject constituted by the legal discourse is impure, enticing, and seductive and thus there is a need for control and surveillance. This control and surveillance is manifested in many practices that have emerged during this period. In the space below, I discuss them in the context of dressing practices of women. I also discuss the notion of *chadar* and *chardewari* as a form of panopticon aimed at discipline and surveillance.

### Chadar[12] and Chardewari[13]: The Panopticon of Dress and Home

The issue of how women should dress is not new to Pakistani society. One need only point to the refusal of Ghulam Mohammad and the *ulema* to sit with unveiled women in the Constituent Assembly. Nor has it been confined to women solely. Zulfiqar Ali Bhutto, for instance, promoted *shalwar kamiz* as the national dress. He also made mandatory a Mao Tse Tung–style uniform for the members of the ruling party, especially for his inner cabinet. Indeed much of his dress campaign was clearly directed at men. Both men and women were encouraged to wear the *awami* (of the masses) dress but were not obliged to. Women were free to wear other dresses, and veiling was expected but not coerced (Rousse 1998: 57).

Women's dress, however, became a major issue during the Zia years. It is interesting to note that the concern momentarily moved away from the issue of *purdah* (the practice of veiling) and was cast in religio-nationalist terms. The nationalist binary of Hindu-Muslim was transposed to women's dress so that *sari* was designated as "Indian" and thus by implication un-Islamic. The *shalwar kamiz* and *dupatta* were declared the national dress for women. This was ironic because, as Mumtaz and Shaheed note, "between Bangladesh and India, there must be between two or three times as many Muslim women who wear *sari* as those who wear *shalwar kameez*" (1987: 78). What is also ironic is that all Punjabi women in India, whether Hindu or Sikh, wear *shalwar kameez* despite the fact that they are neither Muslim nor Pakistani.

The issue of *chadar* also came back in a different form.[14] This time around it was combined with the notion of *chardewari*. Zia ul Haq used the notion of *taqadus* or *namoos* (sanctity) of *chadar* and *chardewari*. Repeatedly he proclaimed that *he* would ensure that the sanctity of *chadar* and *chardewari* was upheld and safeguarded. Partly this was an implicit (but repeated) reference to the excesses of the senior Bhutto era, especially with respect to women. But mainly the notion of *chadar* and *chardewari* was to provide the panopticon from which not only surveillance could be performed but also female sexuality and body disciplined. Where *chadar* and *chardewari* is perhaps different from Bentham's or Foucault's panopticon is that the surveillance in their case was not necessarily for the purposes of extracting biopower. Or perhaps even more important is the fact that surveillance was performed only by the state or the state agencies. Surveillance of women by means of *chadar* and *chardewari* in Pakistan became the domain of all men in the society.

*Chadar* both created and restricted the personal space of women while the *chardewari* restricted and organized their social and cultural space. Though literally *chardewari* means "four walls" (of the house), the restrictive space created by it was not only confined to the family or domestic sphere (space). This space was reproduced outside the home as well. Outside the domestic context, the *chardewari* extended to separate banking outlets for women, separate police stations, separate work space within offices, and the women's university. Wherever women had to be allowed beyond the range of the panopticon of *chardewari*, such as in educational institutions and television, *chadar* mandated the restrictive personal space in which she could operate.[15]

Parallel to this process of controlling women's bodies and sexuality was another process of control. In 1984 in Nawabpur, a small village in Southern Punjab, the female family members of a carpenter suspected of having relations with the daughter of the local landlord were molested and paraded naked through the village. The sanctity of *chadar* and *chardewari* was trampled. The impunity with which the crime was perpetrated and the fact that the perpetrators were not apprehended clearly showed that the purpose of *chadar* and *chardewari* is not to protect women from men: its sole purpose is the surveillance and discipline of women.

The space created by *chadar* and *chardewari* marked the boundary between the acceptable and the abhorrent representation and behavior of women. To quote Shahnaz Rousse at length:

Boundary protection and definition—sexual, geographical, political, social, and moral—became more pronounced after 1977. During the Zia period, a powerful alliance emerged discursively and in power terms between the guardians of the state—the military—and the guardians of public morality—segments of the *Ulema* who collaborated with Zia's regime to use the media, the mosque, pulpit, and private vigilantism to assure intimidation and harassment of any and all elements that challenged conventional wisdoms (read: hegemonic ideological formulations). (1998: 59)

As I elaborated earlier, instead of understanding it as an alliance, the notion of religiopoly—a symbiotic merger of the religious and the militaro-nationalistic discourses—offers a much better explanation. With religiopoly as the dominant discourse, other discourses (notably those of media and education) have drawn upon it to a significant extent. I now turn to discussing this.

## Women and the Media Discourse in Pakistan

As I mentioned in chapter 1, the media discourse in Pakistan (print and electronic) has similarly fixed meanings for what it means to be a woman in Pakistan. This fixation of meaning was partly a result of the process of "othering" (construction of the "other"), which works on two levels. On one level, *woman* in the media discourse is constituted as the other of *man*—the "national hero," the guardian of national and ideological frontiers, the provider of subsistence, and so on. In each case, the superior meaning is ascribed to the other of woman, thus relegating her to an inferior status. On the other level, a binary relationship is created at the intrawoman level, i.e., between the "good" and the bad woman; between what Hafeez (1981: 224) terms as *chiragh e khana* (literally "lamp of the home") and *shama e mehfil* (literally, "party illumination"). Here, once again, the first of the two terms in the binary is accorded superiority over the latter.

In this respect, the media discourse in Pakistan draws upon similar meaning fixation in both classic and contemporary Urdu literature. In literature, the characterization of women, such as by Akbar Illahabadi in poetry and Deputy Nazir Ahmad in prose, creates the good woman/bad woman binary. Take for instance these verses from Akbar Illahabadi's poetry:

> "When I asked what happened to women's veil
> They said, it covers men's thinking abilities now."

> "Hamida did not shine until she learnt her English
> Now she is a party's illumination; formerly she was
>    the household lamp."[16]

Similarly, the main characters in Deputy Nazir Ahmad's famous two-part novel *Binat ul Naash, Marat ul Uroos*, named Akbari and Asghari, also conform to the good woman/bad woman binary. The binary created in the names Akbari and Asghari extends all the way to characterizing the domesticated (Asghari) and the modern (Akbari). In media images, just as in literature, only the *chiragh e khana* can have all the virtues approved by religion and society. In print media, however, especially since the 1980s, the nomenclature used to depict the *shama e mehfil* has also included *wahisha*, *bud kirdar* (both meaning a woman of loose morals), and *randi* (prostitute).[17]

The results of a survey conducted in 2001 by the United Nations Development Programme (Pakistan) on "portrayal of women in media" reported that during 70 percent of telecast time, viewers on the basis of telecasted images perceived women as self-sacrificing and with no identity of their own. Furthermore, the same survey showed that 79 percent of Pakistan Television (PTV) dramas project images of working women as alienated from their husbands and children. According to the survey report, around 70 percent of the time professional women are identified as fashion models, teachers, and doctors, whereas 90 percent of the time men are projected as politicians, national heroes, and government officials (UNDP 2001).

Censorship and state intervention have always been a fact of life for the Pakistani media. However, prior to the late 1970s, censorship was focused more or less on the political content that the state managers thought would be politically harmful to the state. The 1980s saw an extension of censorship to the social realm. As argued earlier, this was the time when religiopoly became the dominant discourse in Pakistan. In 1980, for instance, directives from the government asked PTV authorities to ensure that all announcers and newscasters cover their heads. PTV authorities were also directed to curtail advertisements that had female models or that contained what the government described as "obscenity." In a way, this was tantamount to equating women with obscenity per se (Mumtaz and Shaheed 1987; Hussain and Shah 1991). Furthermore, the directors of television plays were directed not to show any scenes in which a man and a woman sat on the same bed, even if they played husband and wife.

Media texts (plays, advertisements, talk shows, lectures, etc.) drew upon religiopoly to fix meanings for different genders in Pakistani society. Since the scope of this chapter does not allow for a detailed examination of this important discourse in constructing subjects and subjectivities, let me just point to some of the ways in which this took place. First, a different genre of texts started to displace the earlier media texts. Playwrights such as the husband and wife team of Ashfaq Ahmad and Bano Qudsia, with their very fatalistic view of self and society, started to occupy more and more space on the airwaves. On the other hand, plays that projected the feudal culture also began to be aired.[18] Together these texts conveyed an image of the society that was fatalistic as well as traditional. In doing so they portrayed the image of women in essentialist terms.[19] Hussain and Shah's survey of programming during this time included the following essentialist categories in which good women were portrayed: self-sacrificing, self-abnegating, virtuous, domesticated, mother, daughter, sister, honest in poverty, loyal, religious, emotional, irrational. (1991: 169–73). Anyone outside these categories was a bad woman.

The second important way in which the media discourse engages in a meaning fixation exercise is what I term the *hyperreality of the sermon*. Religious scholars do not confine themselves to addressing captive gatherings of men (in the mosque): their message is now beamed right into the living room. Dr. Israr Ahmad (an obscurantist physician turned religious scholar)[20] for years spoke to the whole family over the airwaves. It was not that *ulema* had never before been featured on the television, but the message that was beamed into the living rooms was different. A brief description of Dr. Israr Ahmad's message is necessary to make the point. On women, his stated position was that all working women should be retired and pensioned and that women should not be allowed to go out of the house except in emergencies (*Jang* Daily, March 18, 1982). He went so far to state that "no one could be punished for assaulting or raping a woman until an Islamic society had been created" (Mumtaz and Shaheed 1987: 84).

Though his program was taken off the air after a spirited protest by women, the message remained. Veena Hayat's case[21] shows how the female subject constituted by the media discourse still suffered nine years after *Al Huda* (Dr. Israr Ahmad's TV spot) was taken off air. The print media reported his statements and the *Pakistan Times* (the state-owned daily newspaper) even reproduced his Friday sermons. In the print media during this time, women were not totally absent but were inserted differentially. Politically and socially active women were either totally ignored or were reported sensationally.[22] The vast majority of women in Pakistan only make news when something bad happens to them, be it the Nawabpur case (discussed earlier in this chapter) or the Zainab Noor case.[23] Such fixation of meanings is also evident in androcentric films that reify the panopticon of *chadar* and *chardewari* in the terms discussed above.

## Women's Agency in Pakistan

Even in the face of such absence as discussed above or the differential insertion of women in the media and other discourses, Pakistani women are not without agency. It would not be very far fetched to say that, had it not been for women in the WAF and other such groups, democracy might have been restored in Pakistan much later than 1988. There is no doubt that women were the only group that stood up to the despotic military regime of General Zia ul Haq. Similarly, as Shaheed (1998) shows for Pakistan, agency at the local levels (villages, small towns, etc.) might not be the same as articulated by the feminist theoreticians or activists in metropolitan cities, but it is indeed present and is manifested in mundane act of day-to-day living. As Patricia Jeffery puts it:

> Agency is not wholly encompassed by political activism. Women outside the ambit of high profile activist organizations—whether feminist or not—are by no means passive victims, so successfully socialized into obedience that they cannot discern gender inequalities. In various low-profile ways, women critique their subordination and resist controls over them—in personal reminiscences or songs, in sabotage or cheating. The husband treated like a lord or deity to his face may be derided behind his back or given excessively salty meals. (1998: 222)

Although much attention has been paid in the literature to Pakistani women's agency and activism in the Zia period (1977–1988), their agency is not limited to this period alone. Both before 1977 and after 1988, there are numerous examples of women's agency and resistance against oppressions of all kind. Second, scholars have focused more on the agency and activism of liberal women and women's organization while paying less attention to the agency of women outside this stratum, for example, working-class and rural women. Mumtaz's research (1996) corrects this to an extent by bringing into focus women's organizations and movements in rural as well as urban Sindh, namely the Sindhiani Tehrik and the Mohajir Qaumi Movement (MQM) Women's wing, respectively.

Even less attention seems to have been paid to the phenomenon of right-wing political and social women's groups, such as the women's wing of Jamaat e Islami or Al Huda.[24] With the exception of Jamal's rigorous and excellent analysis (2002) of Al Huda, which looks at it from the perspective of transnationalism, most analysts have not been able to explain the presence and agency of right-wing women. The most common explanations have interpreted them as a ploy of the Jamaat e Islami or other religio-political organizations, or as women who have been so socialized by the Islamization they have internalized that they cannot see the oppression in it. These explanations, however, are unable to explain why some women willingly submit to exploitation and oppression while others do not, or some women internalize Islamization and others do not.

I argue that this inability of feminist literature to explain the phenomenon of right-wing women is due largely to the binaries on which analyses have been based. It is also partly due to the assumptions of woman as a unitary entity. Let me briefly explain. The binary of state/religion that has been used to cast the debate on development has also been transposed to the study of the state, women, and obscurantist forces. It has been assumed that a modernizing state will always be a natural ally of women against the obscurantist forces (religio-political parties, etc.). In consonance with the unitary notion of woman, it is assumed that all women will react to and/or resist religio-political forces in the same way. Thus, not only are feminist scholars surprised at the activism and agency of right-wing women, their frameworks are unable to explain this phenomenon. The fact of the matter is that the religiopolistic discourses in Pakistan constitute different subjectivities and multiple identities. The subjects constituted by the discourses are also positioned differentially. While some of them resist subject positioning in the form of activism (liberal women), others find agency from diametrically opposing positions. Thus poststructuralist feminist perspective, by moving the focus from polar dichotomous binaries and assumption of unitary subjects and identity to multiple subjects and subjectivities, are better able to explain the agency of right-wing women in Pakistan.

Coming back to this point, the Pakistani woman is more than the oppressed, domesticated, exploited soul or the sacrificing goddess. She has agency that she uses differentially to her advantage, from writing songs and participating in the processions during the independence movement to demanding the implementation of the *Shariat* act in 1948 (to get inheritance rights), to facing police baton charges and jails (this time to protest against the oppressive Islamization of General Zia ul Haq), to operating upscale religious schools such as

Al Huda. While this agency has not won for them the rights that they aimed to achieve throughout the history of Pakistan, it is not because they have been passive subjects in the face of oppression. Historically, their agency has manifested in multiple identities, and they have made contingent alliances with different groups (with the modernizing state during the 1950s–1970s, with ethnic groups and movements, MRD, MQM, Sindhi nationalists from 1977 onwards, with minorities 1977–1990, and with right-wing organizations from the 1980s onwards). Subjectivity constitution and agency thus go hand in hand.

In the next chapter I explore in detail how gendered subjects and subjectivities are constituted by the educational discourse in Pakistan, especially in and through curricula and textbooks, and at what points can we see agency and resistance to this subject constitution.

# Chapter 6

# Educational Discourse and the Constitution of Gendered Subjectivities in Pakistan

In this chapter I discuss the constitution of gendered subjects and subjectivity in Pakistan's educational discourse. I especially focus on the mechanisms embedded in curricula and the textbooks through which subjects[1] and subjectivities are constituted and ordered. Curricula, especially textbooks, are the sites where educational discourse, drawing on the economic (peripheral capitalism, global capitalism), political (statist, nationalist) and religious discourses (among other discourses) constitutes gendered subjects and subjectivities. Textbooks are also the sites where these and other discourses compete and contend to assign or fix meanings to different signs in the discourse. Meaning fixation involves ascribing certain values to certain signs (notions, words, symbols, concepts) and the privileging of certain signs by ascribing to them superior meanings. Meaning fixation also involves the use of binaries (e.g., male-female, nation-enemy, etc.) to create the "other" where the first element is good, better, superior, rational, privileged, in other words, the "self," and the second bad, worse than, inferior, irrational, in other words, the "other."

The texts constructed by the discourse constitute subjects. These subjects are ordered, or as Foucault (1979) terms it, disciplined through techniques of normalization, homogenization, and classification. This constitution and disciplining of the subject is different from the traditional notions of discipline through juridical statutes, penal institutions and mechanisms, and systems of punishment and remuneration. As Foucault tells us, "Whereas the juridical systems define juridical subjects according to universal norms, the disciplines characterize, classify, specialize; they distribute along a scale, around a norm, hierarchize individuals in relation to one another and, if necessary, disqualify and invalidate" (1979: 223). The crucial difference is that while the juridical system defines the unitary subject (e.g., in relation to law, the state), the discourse, through discipline, constitutes subjects that are defined in relation to one another. Discourse creates power and power relations. The linchpin of the exercise of power and the regulation of power relations between subjects is surveillance. Here once again the Foucauldian notion of surveillance is different from the traditional notion of surveillance. Surveillance in the Foucauldian sense is not carried out by a single source from a single point in society. It is carried out or exercised from multiple points through the power relations among the subjects. Surveillance, as Foucault (1979) suggests, perfects the exercise of power. For example, the educational discourse in Pakistan constitutes subjectivities in such a way that some subjects are accorded the status (and power) of surveillance over other subjects. Let me use a lengthy quotation from Foucault's *Discipline and Punish* to make this point:

> [Discipline] in each of its applications makes it possible to perfect the exercise of power. It does this in several ways: because it can reduce the number of those who exercise it, while increasing the number of those on whom it is exercised. Because it is possible to intervene at any moment and because the constant pressure acts even before the offence, mistake or crimes have been committed. Because, in these conditions, its strength is that it never intervenes, it is exercised spontaneously and without noise, it constitutes a mechanism where effects follow from one another....In short, it arranges things in such a way that the exercise of power is not added from outside, like a rigid, heavy constraint, to the function it invents, but it is so subtly present in them as to increase their efficiency by itself increasing its own points of contact." (1979: 206)

The distribution of surveillance at various points in society constitutes a gaze that is omnipresent and effective. An example from Pakistan elaborates this point. Educational discourse in Pakistan, and especially the curricula and textbooks, constitute gendered subjects in such a way (as I discuss below) that the male subjects are not only privileged over the female subjects but some of the former are also accorded a status from which they keep a continuous surveillance over the latter. There is thus a constant gaze, which orders the gender (and power) relations in the society. In Foucauldian terms, each gaze forms "a part of the overall functioning of power" (Foucault 1979: 171).

Discipline through the gaze results in the constitution of docile bodies (subjects) normalized into believing in the normality of power and existing power relations. Such docility and normalization is directly linked to the economic, political, and religious functions of the system. As Foucault puts it, the gaze performs "the administrative functions of management, the policing functions of surveillance, the economic functions of control and checking, the religious functions of encouraging obedience and work" (1979: 173–74). In Pakistan, educational discourse is the foremost site where subjects are constituted and disciplined by means of the surveillance gaze.

My main focus in this chapter is on how such subjects and subjectivities are created by educational discourse in Pakistan, through what mechanisms and processes some are privileged over others, how some subjects are ascribed the role of surveillance, and finally, what effects all these have on gender relations in Pakistani society. As I mentioned earlier, there are many processes through which educational discourse in Pakistan constitutes subjects and their subjectivity. One of the foremost in this respect is the process of inclusion and exclusion. Subjects come to be understood by means of their inclusion in (or exclusion from) the dominant meaning fixed by the discourse. For example, by equating "Pakistani citizen" with "Muslim," the discourse (curricula and textbooks in our case) by and large exclude all religious minorities from the meaning of "citizen." Similarly, through other meaning fixation (as I show later), others such as women or linguistic minorities are also excluded from citizenship. What this essentially means is that, by including few and excluding some, the society is ordered in such a way that some are more privileged than others.

Another mechanism through which subjects are constituted is through hierarchization. In this case, the meanings are fixed in such a hierarchical manner where the signs (words, notions, etc., whose meanings are fixed discursively) that are higher up in the hierarchy are understood to be superior to those lower in the hierarchy. Just as the use of binaries by the discourse creates and regulates power relations along the horizontal dimension, hierarchization creates and regulates them along the vertical dimension. The third major process through which subjects are constituted by the discourse is that of totalization, i.e., specification of totality as a means of homogenizing explanation as well as theory. In totalization, the meanings of signs are fixed in a way that they subsume all differences. For example, the notion of "nation" as articulated in the curricula and textbooks in Pakistan obfuscates all ethnic, linguistic, regional, religious, and gender differences. In this exercise of meaning fixation, some signs come to be regarded as normal, i.e., their meaning as understood by the subjects seems given and they are thus never contested.

It is important to understand that meaning fixation not only constitutes subjects but also regulates their behavior and relations between them. This regulation is principally carried out by means of discipline and surveillance. The discourses create mental, intellectual, and physical spaces, which facilitate disciplining, and surveillance of the bodies of the subjects. This is not always explicit and not always carried out by the state or the state agencies. It is instead carried out through intersubject relationships and by the state in relation to these relationships. The knowledge constructed (and imparted) by and through the discourse cannot be separated from power. In other words, the way subjects are constituted and the way they are ordered and disciplined creates power that is delegated to some of the subjects who perform the surveillance tasks. Thus, just as power is not concentrated in one center or source, the means and agents of surveillance are also spread throughout the society. For example, in Pakistan the task of surveillance of women is discursively accorded to all male members of the society. Similarly, through laws such as the blasphemy law, the task of surveillance over religious minorities (e.g., Christians, Hindus, etc.) is accorded to the male members of the majority religious group, i.e., Muslims.

Together, discipline and surveillance order subjects in such a way in the society that their bodies (biopower) can be utilized to the maximum for economic, political, administrative, military, and religious purposes. As Foucault writes, "Side by side the major technology of the telescope, the lens and the light beam, which were the integral part of the new physics and cosmology, there were minor techniques of multiple and intersecting observations. Of eyes that must see without being seen; using techniques of subjection and methods of exploitation, an obscure art of light and the visible was secretly preparing a new knowledge of man" (1979: 171). The importance of educational discourse lies in the fact that it not only constitutes subjects but it also provides the "gaze," the basis of discipline and surveillance.

Each discourse fixes the meanings of signs around a number of nodal points. In Pakistan's case, these nodes, as I discussed earlier, are religion (Islam) and nation (Pakistan as a nation). The meanings of a majority of these signs are derived from and fixed around these nodes. These nodes are not static or fixed and continue shifting in their shape, manifestation, and influence. For instance, while Islam as a nodal point was manifested in the notion of the two-nation theory (TNT) before, during, and just after the independence movement, it shifted to take the form of modernizing Islam during Ayub Khan's era (1958–1969), Islamic socialism during Zulfiqar Ali Bhutto's time (1970–1977), the Nizam e Mustafa and Islamization during Zia ul Haq's tyrannical reign (1977–1988), and the *Shariat* Bill and Talibanization in the 1990s (under the Benazir Bhutto and Nawaz Sharif regimes). Similarly, the nodal point of nation (nationalism, patriotism, *Pakistaniat*) has also shifted a number of times, even up to a point where they merged into a symbiotic interdiscursivity.

Around these nodes of Islam and nation or nationalism, meaning fixation takes place that eventually constitutes subjects and subjectivities. A poststructural feminist reading of these nodes shows that both these nodes are gendered to begin with. Thus, all meaning fixation that takes place around them is also gendered. In the following discussion, I first demonstrate the gendered nature of these nodes. Then I show how the social studies and Urdu curricula and textbooks (texts) fix the meanings of different signs in a gendered manner around these nodes through normalization, totalization, and classification. In doing so, the texts constitute subjects as well as technologies of discipline and control through which the gender and power relations are regulated in Pakistan. Finally, I argue that the female subjects constituted by the discourse are at the losing end of the gender and power relations. Thus education, instead of empowering, actually disempowers women.

## Urdu and Social Studies Curricula and Textbooks in Pakistan: An Overview

For the majority of students who study at the public schools in Pakistan, textbooks are the only source of knowledge. The prescribed textbooks are developed on the basis of curricula approved by the Curriculum Wing of Pakistan's Ministry of Education. These books are often badly designed and badly produced. The quality of research leaves much to be desired. The data are inaccurate, and editorial mistakes abound. Of the two subjects examined for this research, Urdu is a compulsory subject that is taught at all levels starting in class 1, while social studies is compulsory beginning in class 3. Prior to 1958, the curricula contained separate subjects of history, geography, and civics. However, the military regime of Ayub Khan abolished history as a subject and introduced two new subjects: *muasharati uloom* or social studies (for classes 3–8) and *mutala Pakistan* or Pakistan studies for classes 9–12. Both these subjects are an amalgam of history, economics, civics, and social studies (Aziz 1993).

The combination of history, geography, and civics into one subject amounts to a fusing of time, space, and the relations between citizens and the state into one subject of study through which knowledge is to be imparted to the students (Saigol 1995: 208). A close look at the curriculum documents (CD) issued by the Curriculum Wing (CW) and the textbooks that are prepared according to these documents shows that the amalgamation was not really an attempt to provide a multidisciplinary perspective to the students. Each area within both the CD and the textbooks is tightly compartmentalized, and anything that does not fit anywhere else seems to be thrown in at random (Saigol 1995). There also seems to be no attempt to provide the epistemological explanation of amalgamating these disciplines together. No explanation of interlinkages or underlying factors or themes that (might) unite these areas is provided either. These texts both create and blur disciplinary boundaries simultaneously with what Saigol terms "a fragmented view of social reality...[that] produces...violent consciousness" (1995: 208).

Furthermore, both the social studies and Urdu texts (both CD and textbooks) are heavily gendered with a pronounced androcentric bias. There is an equally pronounced bias in favor of males in the authorship of the textbooks and the drafting of the pertinent CD. One pioneering study that looks at the representation of women in school textbooks has noted that in the books surveyed for the study a total of "3819 characters were portrayed in the textbooks...of which 81% were male and remaining 19% were female. In this way the sex ratio came to be 419 males per 100 females which was way up compared to the national sex ratio of 111" (Anwar 1982: 12). Though this particular study looked at gender representation from a quantitative perspective, it came up with some startling findings and provided the basis from which qualitative research on gender representation could be conducted. Anwar demonstrated through statistical analysis that "the greatest discrimination against females was seen in the secondary school textbooks—the level of education beyond which majority of females do not go" (1982: 20). This is incidentally also the age where consciousness of self (and other) is formulated. The study also found that there is a predominance of urban characters in the textbooks (whereas 70 percent of Pakistanis live in rural areas) (1982: 21); that the most frequently occurring activities depicted in the textbooks are what can be termed male activities (i.e., fighting wars, spreading religion, political activities, scholarly activity—especially writing novels, articles, etc.—outdoor hobbies, welfare activities, and religious practices). The most common activities in which females are depicted in the textbooks are cooking, cleaning, childrearing, domestic help, cotton and fruit picking, and cloth and dish washing (1982: 39).

In terms of the images of males and females in the textbooks and their attributes, Anwar reports that "of all the qualities attributed to male and female characters, brave, rational, humane, respectable, cooperative, loving, clever, religious, active, advisor, industrious and responsible were the most frequently mentioned ones in a descending order.... fewer women than men were depicted as possessing these qualities as compared to this [*sic*] women as portrayed in the textbooks...were the least likely to be learned, freedom fighters, leaders, patriots, rebels and genius." The only attribute in which women beat the men in terms of representation is in "being domesticated" (1982: 46).

While the study is commended for whistle blowing—and this too under the tyrannical regime of Zia ul Haq when such activities were considered subversive and even un-Islamic—it suffers from two major shortcomings. First, it looks at gender categories in unitary, singular terms. In other words, it proceeds from the assumption that man and woman are the only two analytical categories available to examine gender balance in textbook representation. Thus, by default it overlooks various sexual orientations and multiple identities of both men and women in the society.

Second, by focusing on the quantitative characterization of males and females in textbooks, the study falls into the positivistic trap of assuming that, unless there is a visible characterization (of males and females), there cannot be any discrimination. As Anwar states in the beginning of the study, one of its guiding assumptions was that there is no gender bias in science textbooks since they are not "expected to depict social situations and [were] likely to contain only few human characters" (1982: 7); hence, these were thus excluded from the study. Feminist scholarship has amply challenged this line of argument and has demonstrated the gendered nature of science itself (Haraway 1991; Harding 1990).

Finally, the study does not analyze the social implications of this lopsided characterization of females in the textbooks. This conveys an impression that the only problem is one of textual and graphic representation and that once this problem is addressed and solved, the gender imbalance can be corrected. This, however, has not proved to be the case. Both the textual as well as graphic representation of females has improved in textbooks since Anwar's study (e.g., in Urdu textbook class 3, the female graphic representation is now at the ratio of 1:2; similarly, textual representation has also improved), and yet socially the condition of women has gone from bad to worse in the years since the study. While ideologically standing with Anwar (1982), I shift the focus of inquiry into gender imbalance by asking: why is this imbalance present in the textbooks in the first place? Furthermore, what consequences does this imbalance have for the larger society, especially in terms of the production and reproduction of gender and power relations and the relationship of these relations to the state?

## Nodal Points: Islam and Nationalism

As mentioned earlier, religion has been a nodal point of political and educational discourse in Pakistan since the early days of the Pakistan movement (1930 onwards). Prior to this, Islam was never really an issue around which meanings of other signs such as nation, community, or identity were fixed. This is not to say that Muslim identity was not there at all prior to the demand for a separate state by the Muslim League elite late in the movement for independence from British rule in India. As I have elaborated earlier (chapter 4), following Partha Chatterjee (1993), it was the Orientalist discourse of the British colonial project that gave it a communal meaning by understanding and dealing with Hindu-Muslim differences as communal antagonism based on religion. Once this was done and once the nationalist leadership, primarily that of the Muslim League in this context (and that of the Indian National Congress in the larger context of the independence movement), started to articulate political ideas and the process in Orientalist terms, only then did Islam become a nodal point of Muslim politics in India.

Also as I have discussed earlier, the colonial cultural, political, and juridical intervention in the respective Indian systems was also instrumental in bringing forth religion (Islam) as a marker of identities for Hindus and Muslims of India. Since then, however, it has remained as a nodal point of Muslim and later Pakistani politics even though it has shifted in form and nature over time in accordance with the competition and contention between different discourses, especially with the modernizing discourse of the nationalist leadership (prior to independence) and the modernizing nationalist state after 1947. Then, as argued earlier, the two discourses, while maintaining separate discursive and articulatory domains, fused by the late 1970s into a symbiotic relationship for which I have coined the term *religiopoly.* In the following sections, I show how Islam has, over time, acted as a nodal point around which the educational discourse in Pakistan has articulated gender and gender relations (in and through curricula and textbooks).

By the 1930s, the nationalist ideology and political demands of the Muslims of India represented by the Muslim League (which claimed to be the sole representative of all Muslims of India) were defined around religion and came to be articulated as the two-nation theory (TNT). The TNT offered the Muslims of India an idea of nationalism that, unlike the European notion of nationalism as a substitute for religion, was grounded in and derived from religion. Simply put, the TNT put across the idea that Hindus and Muslims formed two separate nations (not two groups within the same nation) that, come what may, could never live together in harmony. Thus there was a need for a separate homeland for the Muslims of India once the British left the subcontinent. The articulation of nationalist ideology and political demands in religious terms had two important consequences for the articulation of gender in Pakistan's educational and other discourses. First, ideology articulated in religious terms not only created a binary by dichotomizing Hindus and Muslims into two distinct (nations) groups but also articulated the Hindu as the "other" of the Muslim "self," with each representing a diametrically opposite worldview.

Second, the TNT, by articulating the Muslims of India as belonging to one nation (as against the Hindu other), laid the foundation for a theoretical and explanatory unity that did not have the sophistication to look at or deal with heterogeneity in the form of ethnic, religious, linguistic, and gender differences within the "oneness of nation." The discursive influence of the articulations around the nodal point of the TNT did not stop at independence and the separation of Muslims from Hindus. It carried on, and in the same process affected, shaped, and constituted all social relations. The religious discourse defined and constituted particular and peculiar subjects who, in relation to each other, were to reshape the discourse itself. It also effected social relations between men and women, between the majority and minorities, and between ethnic, linguistic, and religious groups that were to define the state and governmentality in relation to subjects.

Of the many forms of social relations, gender relations are of special importance because the TNT as a nodal point was essentially gendered in nature. The subjects that are constituted by the education discourse are gendered and so are the relations among these subjects. Here it is important to unfold what I mean by *gendered subjects* and *subjectivities*. On one level, subjects are gendered in terms of the male-female binary, where the first sign is ascribed a superior meaning over the latter. This genderedness is not, however, fixed or immutable. At another level, several subjects, both male and female, are positioned in the discourse whereby they have a feminine subjectivity, especially when in their relationship with the state. At this level, the state is discursively gendered with masculinity while the subjects in relation to it are gendered with femininity. On the third level, the othering is also gendered. The "other" (e.g., Hindu, Jew, nonpatriotic Pakistani, antistate, anti-Islam, non-Muslim, antimilitary, antinuke) is gendered as feminine. This subject is attributed qualities that are considered by the discourse as feminine (cunning, untrustworthy, devious, seductive, sinful, etc.).[2] On the fourth level, time and space are also gendered (Saigol 1995). The pre-Islamic and pre-Pakistan era (history) is feminine and thus dark, sinful, ignorant, and so on. The space (geography) is gendered in such terms as motherland, honor of the mother, fatherland. Finally, in relation to the military, with its phallic symbols such as missiles and warheads,[3] the nation is feminine and thus in need of protection for the sake of its honor. Thus, both the state and the subjects in relation to the state and one another are gendered in a bisexual manner (Saigol 1995).

Coming back to the TNT as a nodal point, three speeches of M. A. Jinnah are extensively used in the Urdu and social studies texts. The first in this respect is his address on March 23, 1940, in Lahore (better known as the Pakistan Resolution)[4] in which, according to most textbooks, he demanded a separate state for the Muslims of India. An Urdu textbook for class 7, for example, states that "Quaid I Azam announced (*Ailan farmaya*) in the 23 March 1940 session that 'No matter how nation is defined, Muslims by every definition are a separate nation and they deserve to have a republic and a state of their own' " (Punjab Textbook Board [PTB] 2002f: 151). The text then goes on to quote another of Jinnah's "famous" addresses(Aligarh, March 1944) in which he said:

Pakistan came into being the day when the first Hindu was converted to Islam. This was when there wasn't even a Muslim state in India. The basis of Muslim nationalism is *Kalima e Tauheed* and not the state or race. When the first person in Hindustan converted to Islam he did not remain a member of the former nation. Rather he became a member of a different nation. Thus a new nation in India came into existence. (PTB 2002f: 152: Also see PTB n.d.d.: 24)

Although the texts of Jinnah speeches themselves indulge in binary construction and classification, what is even more interesting to note is the textual interpretation provided for these speeches in the textbooks. After citing these excerpts, the aforementioned textbook imparts the following analysis to class 7 students:

He [Jinnah] explained the reality that the Muslims of the sub-continent are a separate and 'permanent' (*Mustaqil*) nation thus Muslims and Hindus can never become one nation. The need for this explanation arises because the Congress wanted to convince the English rulers that the people of India are one nation and it (Congress) is the representative party of them all.... Though Congress claimed to be the representative of all Indians in reality it only worked for the interests of Hindus and it was quite possible that after getting the government it would have made the Muslims slaves. Quaid was fully aware of the devious nature of the Congress. (PTB 2000f: 151)

The Urdu textbook for class 7 under the 2005 reform (PTB 2009: 69) presents a similar articulation. The text cites a 1943 speech of Jinnah in which he told the Muslim Girls Federation that the Muslims of India tried for 25 years to have better relations with Hindus. However, the possibility of a good relationship was never there to begin with as Muslims and Hindus had nothing in common. Hindustan was never one country geographically, culturally, and politically. It was never inhibited by a "one" nation.

Here, the text does two things. One, it constructs the Hindu as an "other" of the Muslim. Two, it articulates both the self and the other in gendered terms. Anyone aware of language usage in the subcontinent would recognize the usage of feminine attributes for the Hindu. In the above-cited excerpt, Hindus are articulated as conniving and devious—both attributes that are generally reserved and used for women. Jinnah, on the other hand, is attributed with wisdom and insight (he knew what the Hindus were up to)—both being qualities that are generally considered to be masculine (see for instance PTB 2002c: 38: "Quaid i Azam was resolute, firm and brave"; PTB 2002d: 7–8, Jinnah's love for knowledge and Islam; PTB 2002e: 31–32, the leader and the visionary, the uniter of Muslims). Interestingly, as mentioned earlier, this gendering is not fixed and shifts according to the subject and the subjectivity that the text is focusing on. Take for example the essay in the same Urdu textbook for class 7 students, which narrates the story of "*Dushman Hawabaz*" (The Enemy Pilot). The narrative, after telling the story of an Indian pilot who crash lands in Pakistan after being shot down, paints a picture of Pakistanis (Muslims) as kind-hearted and conscientious and who take extraordinarily good care of the injured pilot. The story ends with the Hindu pilot wondering to himself how wrong the information was that he had learnt all his life:

> Never to have mercy on Muslims, always play mischief with the neighboring country, and weaken them to an extent where they can't even think about freedom....He remembered that Hindus sacrificed people from other religions on the altar of Kali, and considered everyone untouchable but themselves. (PTB 2002f: 221)

Here we can see a shift in the gendering by the text. On a moral plane, the gendering is reversed. Muslims have feminine qualities of kindness, love for humanity, tolerance, gentleness, and mercy whereas the Hindu has the masculine attributes of cruelty, injustice, savagery, and violence.[5] This contingent bisexual gendering of self and other is present in almost all Urdu and social studies textbooks for classes 1–8. It is around this nodal point of religion (initially manifested in TNT) that a gendered and fragmented exercise of meaning fixation is carried out in the educational discourse in Pakistan.

In other words, the discourse articulates and people come to understand self and other and their relationship with one another and with the state in highly gendered and nationalist terms. Due to the nature of the pedagogical system in Pakistan, where questioning is not permitted and where the authority of the text and the teacher (the knowledge provider) is immutable and final, this gendered reality is presented as the absolute truth (see Hasnain and Nayyar 1997; Waseem 2001). The juxtaposition of nationalism with religion (and Hindu with *kafir*, "infidel") results in a situation where everything that the text says has the authority and sanction of religion. Conversely, anyone who questions this is declared anti-Islam and thus liable to be disciplined and punished.

There are three major ways in which meaning fixation around the nodal points of religion and nationalism takes place. However, I must point out that these are not the *only* ways in which the meanings are articulated. I also must mention that the separation of these three is purely arbitrary and done for the purpose of organizing the argument. There are of course areas in which these overlap, are interlinked, and draw upon each other. These include: totalization, whereby meanings are fixed to achieve the specification of totality as a means of homogenizing explanation and theory; classification, through which a hierarchy of meanings is assigned to different signs just as a hierarchy of relations within the society is achieved; and normalization, whereby the meanings assigned to different signs, subject positions within the discourse, and the subjectivities constituted by the discourse are normalized (made to be understood as normal, natural, given, and thus incontestable).

## Meaning Fixation in the Educational Discourse in Pakistan

### Homogenizing Theory, Explanation, and Reality

One of the major ways in which meaning fixation takes place and subjects are constituted and positioned within the discourse is through an across-the-board homogenization of theory and explanation. A prime example of this is the use of the totalizing notion of nation in the texts. Theoretically, such totalization is justified by drawing upon the religious notion of *ummah* (the global Muslim community). The notion, however, is presented in a way that is both divisive and totalizing at the same time. It is divisive in that it separates Muslims from non-Muslims and totalizing in the sense that it obfuscates all cultural, linguistic, sectarian, and linguistic differences within the larger Muslim community. On the level of explanation, this totalization seeks to subsume all differences in the notion of national unity and nationalism. Anything that is not or cannot be made a part of the theoretical or explanatory unity of meaning is pushed out into what Laclau and Mouffe (2001: 111) call the "field of discursivity."

Urdu and social studies texts in Pakistan (CD and textbooks) strive to achieve this theoretical and explanatory unity in a number of ways. The curriculum document for classes 1–3 (Government of Pakistan 2002c; hereinafter referred to as GoP 2002c) that guides textbook writing and production lays down the following general and specific aims, competencies to be built, and content guidelines. For classes 1–3, one of the general aims (third in the list) is love of and familiarity with the Islamic faith and raising children with Islamic values. The corresponding specific aim is learning of Islamic values. The corresponding (proposed) content according to the CD should be comprised of learning the Arabic recitation of two verses of the Qur'an. Another general aim is to produce love of country and nation with respect to Pakistan. The specific aim corresponding to this overall general aim is to introduce to the child the concept of *watan* (a masculine notion of country). The competency that the CD aims to develop is "love of *Watan*" and the way this should be done is by "explaining the country's name and people in such a way that children start to feel attracted to *Watan*" (GoP 2002c: 4–11).

An earlier CD (GoP 1994, quoted by Hasnain and Nayyar 1997) aimed at building the following competencies in children through the teaching of Urdu as the "national" language:

1. to be able to take pride in the Islamic way of life, and to try to acquire Islamic knowledge and to adopt it;
2. to read religious books in order to understand Quranic teachings;
3. to listen to events from Islamic history, and be able to derive pleasure from them (*khushi mehsoos karen*);
4. to know that national culture is not local culture or local customs, but a culture whose principles have been determined by Islam.

The CD for Urdu for classes 4 and 5 explains the importance of Urdu in the following terms. Internationally, Urdu is one of the most commonly and frequently used languages. As a proof of this, the CD argues that Urdu broadcasts are listened to in most countries of Asia, Africa, and Europe. It further states that during the Pakistan movement, Urdu represented the sentiments of Muslims of the subcontinent and since then has been recognized as the national language of Pakistan. The CD authenticates its claims by stating that Quaid i Azam "stamped his approval on this *reality* in 1948." Thus, Urdu, the CD argues, should be introduced as the cultural, social, and civilizational (*tehzeebi*) language of the Muslims of the subcontinent (GoP 2002k: 1). Apart from the fact that the text is factually incorrect,[6] it not only aims to present a totalizing view of "national," nationalism, and language, but it also articulates the meanings of these terms in a homogenizing way. In constructing a discursive "oneness," it discards all other meanings that could be attributed to these signs and thus closes the door on all forms of diversity.

In Urdu and social studies textbooks, the CD guidelines are followed faithfully. These textbooks are replete with totalizing meaning fixation. The most common technique used is reference to Jinnah's speeches, often without giving the students any of the context in which these speeches were made. Other than this, stories from daily life, historical narratives and poems, and so forth are also replete with totalizing innuendos. Let me cite only a few more obvious examples from the texts to highlight this point. The class 5 Urdu textbook, for instance, tells the students that "Quaid i Azam united the Muslims through his untiring efforts and gave them the lesson of faith, unity and discipline and said, 'We the Muslims believe in one God, one Prophet (Peace be upon him) and one book. Thus, it is mandatory upon us that we should be one as a nation too.... If we start thinking of ourselves as Punjabis, Sindhis, etc. first and Muslim and Pakistanis second, Pakistan will disintegrate'" (PTB 2002d: 10). The Urdu textbook for class 6 quotes Jinnah's March 1948 speech at Dacca to the same effect: "What is the use of saying we are Punjabi, Sindhi or Pathan? No we are Muslims.... you will agree with me that whatever you are, wherever you are first and last you are *Musalman* and you properly belong to a nation"[7] (PTB 2002e: 32). Urdu textbook for class 8 under the new reform provides a similarly homogenizing articulation by citing another excerpt from Jinnah's 1948 address in which he said, "I do not want you to talk in terms of Bengali, Punjabi, Sindhi, Balochi and Pathan.... No, you are Muslims first and last" (PTB n.d.e.: 26).

The homogenization of all diversity and difference into a unitary reality—the *Qaum* or nation—has multiple effects. On the one hand it obscures all difference for the sake of formalizing the imagined community (nation). On the other it asks for all sorts of "sacrifices" from the subject in order to achieve, maintain, and defend this imagined community. Just as the colonial state extracted payments from its colonial subjects for its "civilizing" mission, the postcolonial nationalist state extracts payments for providing security and defense. As Nandy puts it, "They [are] not called payments though. They [are] called sacrifices for the future of one's country" (2003: 8). True security, as the texts articulate, is to be found in the oneness of the nation or *ummah*.

It is interesting to note the metaphors for unity in this meaning fixation exercise. It starts with the notion of one God, one Prophet (PBH), and one book. The unitary metaphors then move on to construct the notion of one leader (the *quaid*), one *qaum* (Pakistani), a unitary source of security and defense (the state and the all powerful military), and so on. Such articulation affects the relation between subjects at many levels. For instance, by obscuring all differences, it produces a fragmented consciousness that is often violent. Furthermore, it also obscures intragroup differences by homogenizing and totalizing identities such as men (as one identity), women (as another unitary identity), and so on.

The cumulative effect of this totalizing articulation can be seen at two levels. One, all groups are expected (if not required or forced) to subsume their identity in the larger group (*qaum*), and two, all individuals are expected to subsume their identities into their respective groups and then the larger group itself. Thus, if the group or the individual loses (or capitulates) some rights, it is justified in the larger interest of the *qaum*. Conversely, any concession given to an individual becomes a concession (it is termed as an "achievement") for the whole group. Thus, Benazir Bhutto's ascent to power or the appointment of Zobaida Jalal as minister for education in the Musharaff regime is hailed (or touted) as an achievement for all women of Pakistan.

Another way in which totality is specified is through homogenization of time in Urdu and social studies texts in Pakistan. Totalization of time is primarily done through a representation of history in which the boundaries of time become fluid and contingent. The first step in this process is the construction of imagined time boundaries[8] in terms of before and after: pre- and post-Islamic (before and after the sixth century with reference to Arabia and before and after the fifteenth century with reference to India); and pre- and postindependence (before and after 1947). This construction of imagined time boundaries is gendered in the sense that the pre- or before is always ascribed feminine attributes and represented as time of darkness, ignorance, sin, moral laxity, war, sexual normlessness, and so on. The post- or after, on the other hand, is ascribed with masculine traits such as light, enlightenment, civility, high morals, and so on. Then the texts liquefy the "pre" (pre-Islamic and preindependence) time boundaries in a binary relationship with the post-boundaries (postadvent of Islam in Arabia and India and postindependence Pakistan) of time. The same gendered construction applies to the fluid boundaries as well. Then the boundaries are coagulated into infinite time by the use of "never," "ever," and "always" (Saigol 1995: 218).

A gendered totalization of time in Urdu and social studies textbooks in Pakistan runs something like this: pre-Islamic Arabia and pre-Islamic India are times of darkness, ignorance, intrigue, war, sin, and so forth, whereas the post-Islamic time in both Arabia and India is marked by light, enlightenment, gallantry, civility, and so on. First by marking and then by collating the boundaries of time, this span of one thousand years between the advent of Islam in Arabia and its advent in India is totalized as one period. The gendered construction of time also then portrays Muslims as saviors who salvaged the people of Arabia and India from the dark ages. At the same time the people of pre-Islamic Arabia and Hindus are also collated as one and the same.

In other words, in the textual representation, Hindus virtually become an extension of the Quraish of Makkah, despite a span of more than one thousand years between them. In the same way, Indian and Pakistani Muslims become an extension of early sixth-century Muslims who, after being oppressed by the "infidels," rose up and fought back and eventually not only won but also salvaged these civilizations. The Indian Muslims' quest for a homeland becomes an extension of the *Hijra* (Prophet Muhammad's migration from Makkah to Madina) and Pakistan itself an extension of Madina; the first city-state of Islam. Thus, not surprisingly, Pakistan is often referred to in the textbooks as the bastion of Islam (*Islam ka Qilla*). Pointing to the gendered nature of the articulation of time in the social studies textbooks, Saigol (1995) points out:

> The most important role in which the Muslim male appears is the role of the conqueror, the liberator and finally the subjugator. The imagery used is of a virgin land in need of moral correction lying in wait for the liberator from the *outside* who *enters* the land from some distant country and inseminates it by conquering with his *sword* and then purifying it with his virtuous ideas. The sexual undertones of this imagery are unmistakable. Land and territory (*Dharti, zameen*) are feminine words in Urdu and control (*Qabu*) is masculine. (1995: 229, emphasis in the original)

The *muasharati uloom* textbook prescribed for class 7 articulates the pre-Islamic time on the Arabian Peninsula in the following terms:

> The period prior to the advent of Islam is called the era of ignorance. During this time the Arabs were at the nadir of their religious, social, political and economic life. The whole society was in a state of ignorance, infidelity, waywardness. They did not know anything about government and the state and the society violated the human values....Idol worship was common and each tribe had its own gods and goddesses. They also prayed to the moon, the stars and wind....Sacrificing human

life to please gods was commonplace....During this time alcohol and dancing was common...women were looked down upon and most people buried their daughters upon birth. (PTB 2002m: 8)

This articulation is also echoed by the social studies textbook for class 6 under the new reform (PTB n.d.f.:, p. 94–101). Another social studies text makes the time boundaries fluid in the following way:

Before Islam, people lived in untold misery all over the world. Those who ruled over the people lived in luxury and were forgetful of the welfare of their people. People believed in superstitions...and worshipped false gods. In Iran and Iraq people worshipped the sun also. In South-Asian region the Brahmans ruled over the destinies of the people. They believed that certain human beings were untouchable. There was an all powerful caste-system. The untouchables lived worse than animals. Human beings were sacrificed at altars of false gods. (PTB 2002k: 13)

As the texts move into the Islamic era (the period after or post-), the articulation of time and subjects starts to change. The time becomes one of stability, enlightenment, positive change, tolerance, and virtue. It was the Muslims (both those in Arabia and especially Muhammad Bin Qasim in India) who changed the darkness into enlightenment. According to a masculinized articulation (PTB n.d.h.: 94–97), the spread of Islam was due to the Muslim conquests, the missionaries, traders, and those Arabs who settled in foreign countries. Incidentally, once again all vocations attributed to Muslims are masculine.

Another social studies text states, "The Muslims did not conquer countries merely to annex lands. In many countries the tyrannical rulers ruled their subjects with a high hand. The people of these countries themselves invited the Muslims to come to their rescue. They considered their salvation to lie in the victory of Muslims because Muslims treated subjects extremely well" (PTB 2002k: 21). Here once again the boundaries of time are made fluid and fuzzy. The reference to the invitation extended to Muslim conquerors is obviously to the Muslims invasion and conquest of India epitomized by figures such as Sher Shah Suri, Mahmood Ghaznavi, and Muhammad Bin Qasim. The case of the last mentioned is interesting to note: not only does the text make a huge jump in terms of time (and space), it also engenders time.

Muhammad Bin Qasim, as the story goes, was sent to India to free a shipload of women who were "enslaved" by the Hindu Raja Dahir.

The imagery created by the text portrays Qasim as a young (17 years old), brave, strong, and capable warrior coming to rescue helpless Muslim women from the clutches of the evil Raja Dahir and then laying the basis for the spread of Islam in the subcontinent through later conquests by Mahmood Ghaznavi, Shahab ud din Ghauri, Qutub ud Din Aibak, and so on. Two points are important to note. One, Qasim and Dahir become the respective prototypes of the Muslim hero and Hindu villain for ever after (see for example PTB 2005: 94–97). As Mubarak Ali (1993: 231) states, "all subsequent representations of Hindus are derived from this initial one." Similarly, all subsequent representations of Muslim saviors are derived from Qasim. An example of this was the self-characterizations of generals Ayub Khan and Zia ul Haq, who saw themselves as modern-day Qasims who saved the *qaum* (feminine) from debauched politicians.[9] Two, the texts totally neglect to mention the women who were enslaved by Dahir and liberated by Qasim (Ali 1993a: 13). The women's usefulness is over, while Qasim remains in the texts as the quintessential savior.[10]

Space is also articulated in totalizing terms in Urdu and social studies texts. Almost the entire social studies textbook for class 7 (PTB 2002k; PTB n.d.i) is devoted to an articulation of space in totalizing terms. The list of chapters includes topics such as land features of the Muslim world, major climatic regions of the Muslim world, resources of the Muslim world, people of the Muslim world, trade of the Muslim world, etc. The contents of the first chapter talk in terms of mountains and seas of the Muslim world as if there were a contiguous, monolithic entity called the Muslim world. The cartographic representations also show an image of a single entity that depicts the Muslim world. What the articulation does is create an imagined space and an imagined map that subsumes all geographical, political, cultural diversity. Coupled with the concept of a Muslim *ummah*, the totalizing picture is complete. It also creates the other.

Once the map of the imagined space is ingrained in the mind of the students, it is easier for them to see the entire world from this reference point. At the same time it is very difficult to accept any alteration in this mental map. Thus, the thought of any incursion in space (real or imaginary) by the "other," whether the latter be Jewish or Hindu or Soviet/Russian or American, ignites the patriotic/nationalist subjectivity. In the next chapter, I discuss how space is also articulated by the text through classification, i.e., the differentiation of groups and the construction of the "other."

# Chapter 7

## Classification, Normalization, and the Construction of the "Other" in Pakistan's Educational Discourse

## The Articulation of the "Other" through Space

Texts use gendered notions of space to classify groups and construct the other, especially with respect to Hindus and Muslims. Foremost in this respect is the designation of the space (country) in gendered metaphors. Commonly used metaphors are motherland, mother, *dharti*, and so on (virtually every Urdu and social studies textbook). Another metaphor commonly used is *dharti ki kokh* ("Earth's womb") from which crops grow, or *arz e watan* (literally "earth of the country"), to obtain which (in 1947) thousands of people (read: men) sacrificed their material belongings or lives and women lost "everything" (*sub kuch*, meaning their honor), while to safeguard this martyrs shed their blood.

Such articulation of space has multiple connotations. First, space articulated in this manner becomes feminine and thus in need of protection: who can better provide this protection than (of course) the masculine state and the military? The possession and safeguard of this space was possible, according to the texts, only because of male resolve and bravery (Jinnah in particular and Indian Muslim men in general). Second, this feminine space is constantly threatened by the enemy[1] (Hindus, Jews: the "other") who constantly casts a *gandi nazar* ("lecherous eye") on the motherland. Anyone familiar with South Asian culture is familiar with the connotation of the "evil eye." Usually petty feuds start on the grounds that someone (male) had cast a *gandi nazar* on the woman or womenfolk of the aggrieved party. It then becomes imperative for the men to avenge and restore the honor of the women and the family.

The second articulation of space involves construction of spatial boundaries in gendered terms. A 1988 social studies text (PTB 1988: 90), for example, constructs a gendered architectural binary between Hindu and Muslim spaces of worship. "We (Muslims) worship in mosques. Our mosques are open, spacious, clean and well lit. They (Hindus) worship in temples (*mandir*) and their temples are extremely narrow, enclosed and dark. Inside these temples the Hindus worship idols." The latest social studies text for class 6 extends this spatial binary to the living space by stating, "The Hindus live in small and dark houses" (PTB 1988: 91). There are two complex, interrelated dynamics of othering at play in this articulation of space. One is the construction of a pre-Oedipal/Oedipal binary while the other is, once again, a feminine gendering of the Hindu by equating the sacred architecture with a woman, especially a menstruating woman[2] (dark, unclean, enclosed) (Saigol 1995). In another sense, such articulation also ties in with the time binary discussed above. The Hindu space is pre-Islamic, whereas the Muslim space is post-Islamic. As argued earlier, the articulation privileges the latter over the former.

## Articulation through Exclusion

One of the most common techniques of articulation or meaning fixation is through inclusion and exclusion. Since subjects understand and relate to each other by means of meanings to which a sign, word, or notion is assigned, only those who fulfill the criteria laid out by the articulation are included in the meaning of that sign. The rest are automatically excluded, or in other words, "othered." Articulation of the other serves to order the subject positioning through classification and hierarchization, as I shall explain later. Such articulation, as we saw earlier, takes place around the nodal points of Islam and nationalism. Having discussed the temporal and spatial othering and binary construction between Hindus and Muslims, I now turn to other dimensions of classificatory articulation by the educational discourse in Pakistan, especially with respect to the exclusion of women and minorities.

With respect to the minorities in Pakistan (religious, linguistic, ethnic), the exclusion is explicit and brazen. A general reading of the texts gives the clear impression that there are no minorities in Pakistan. For instance, in all chapters on the population of Pakistan in social studies texts, the population statistics and explanations of these statistics are given in terms of whole numbers (See PTB, 2002h, 2002i, 2002j, PTB, n.d.f., PTB n.d.g, PTB, n.d.h., PTB n.d.i.). No breakdown of the minority population in Pakistan is ever given. In the explanation, while the Christian minority might get an occasional mention, the Hindu minority never makes an appearance.

Parsis, Bohras, Khojas, Memons, and so on only make it into the texts in the context of their support for the Pakistan movement. The texts even omit to mention that Jinnah himself belonged to the minority Parsi community. Sections on prominent personalities (*mashaheer*) do not include personalities such as justices Cornelius and Dorab Patel, Sir Zafarullah Khan (the first foreign minister of Pakistan, who belonged to the Qadiani aka Ahmadi minority declared non-Muslim by Zulfiqar Ali Bhutto) or even the Nobel laureate Dr. Abdul Salam (also a Qadiani).[3] Similarly, the ethnic, linguistic, and religious minorities belonging to other sects of Islam, such as the Shias or Ismailis, are also by and large excluded by way of not mentioning them at all. The overall picture that the texts paint is that of a homogeneous population.

Equally glaring is the exclusion of Bengalis from the texts on the Pakistan movement and history of Pakistan until 1971 (Hoodbhoy and Nayyar 1985; Aziz 1993; Saigol 1995). This exclusionary articulation wipes out (and I must say quite successfully) Bengal and Bengalis from the national consciousness. In interviews with students, what stood out in this respect was that the students' knowledge of Bengal, East Pakistan, and the Bengalis was limited. They could only articulate about former East Pakistan in terms of Indian aggression and the treachery and betrayal of Bengalis. One male student of class 6 told me that "even in the Bangladesh war it was the Hindus of Bangladesh who helped the Indians to breakup Pakistan" (Field diary, translation mine).[4]

Aziz succinctly analyzes the exclusionary classification of cultural and linguistic minorities by these texts in the following terms:

> [This] arid zone was ungraced by any literary creation, social advance, educational progress or intellectual activity. Baluchi folk poetry and classical stories, Pathan [*sic*] poetry and Pushto literature and Khushal Khan Khattak, Sindhi letters, Islamia College of Peshawar, Sind Madrassa of Karachi, Khudai Khidmatgar's social revolution in the NWFP...all this and much...is hidden. (1993: 44)

A more subtle case of exclusion and inclusion pertains to the role of the *ulema* in the Pakistan movement. It is a well-known fact that the *ulema* during the Pakistan movement were tooth and nail against the creation of Pakistan and justified their opposition on the ground that such demands aimed at dividing the Muslims of India. The social studies and Urdu textbooks, in consonance with the discourse of religiopoly, articulate this role in two ways. While most of the textbooks from the pre-Zia era fail to mention this fact altogether, those written and produced after this period claim that the *ulema* were as much as a part of the Pakistan movement as the nationalist leadership[5] (PTB 2002g: 151–55, n.d.e: 47–48). One text goes so far as to include Maulana Maudoodi and Maulana Mahmud Hasan in the list of founders of Pakistan (*Muasharati Uloom (Lazmi)* (n.d.) cited in Aziz 1993: 88).

The process of classification (through exclusion) of Pakistani women is somewhat different and more complex from the one used for the ethnic, religious, and linguistic minorities. First and foremost there is the visual exclusion of women from the texts. This is accomplished by making women less visible in the images in the textbooks, as mentioned earlier. However, in addition to the visual exclusion, it is the meaning fixation through these images that articulates gender in these texts. Almost all visual representations of women in the textbooks, for example, define the physical space that the women should occupy.

The space is constructed in two ways. First is the psychological space that is marked by what I have termed previously (see chapter 5 the *panopticon of dress*. All graphic representations of women in the social studies and Urdu texts show them wearing *shalwar kamiz* and *dupatta*. Second, most if not all graphic depictions show women in physical spaces that are traditionally and discursively feminine, e.g., home, classroom (as teacher), and hospital (as nurse). Rarely is a woman shown working in a laboratory, field. office, or bank (which a lot of women in Pakistan actually do). The texts thus fix the meaning of space in which the female subject and her subjectivity is ideally to be located. In other words, they position the subject in a particular space outside of which she cannot be understood (or is not meant to be understood). Those women who are forced or those who choose to be outside these spaces face some of the social and legal consequences mentioned in chapter 4.[6]

Another form of women's exclusion from the text is their total absence as a distinct category in the population figures. As in the case of minorities, all textbooks give composite population numbers. No gender breakdown of the population is ever given, thus creating a degendered, homogeneous image of populace of Pakistan. One of the textbook writers that I interviewed wondered if this was a case of the government considering men and women as equals (personal communication). It is amply evident from reading the curriculum document and the textbooks that this is not the case. The texts consider women both different and unequal in relation to men. Difference is evident in the subjects' positioning within the text (and the discourse) and inequality evident in the frequency and nature of their inclusion in the text.

In terms of exclusion, the first type mandated by the discourse is from the body of scholars who frame the curriculum documents and those who write the actual textbooks. In comparison to men, there are far fewer female members of the curriculum committees, expert committees, and female textbook writers. Though I was proudly told that both the Curriculum Wing of the federal Ministry of Education and the Punjab Textbook Board were headed by females, both told me in respective interviews that it had not been easy for them to get to their positions and that it was a struggle each day to gender sensitize the working of their respective institutions.

Furthermore, in cases where the textbook is not written by a team and is composed of selections of pieces written by various authors, poets, and so on, the exclusion is even more glaring. The CD for Urdu (classes 11–12), for instance, suggests that only two pieces written by women authors be included in the textbooks (GoP 2002n: 7–10). In stark contrast to this, the CD recommends 51 pieces by male authors. The document "consciously recommends" (GoP 2002n: 17) that all language textbooks should be on the same pattern. A look at the composition of the committee that drafted this document reveals that, with the sole exception of one woman who is an ex-officio member in her capacity as the director of the Curriculum Wing, all other members are male (GoP 2002n: 20). There are no recommendations made by the CD for participation of female textbook writers in developing Urdu textbooks for classes 5 through 12 (GoP 2002k: l, m, n).

Another way in which the texts use classification is through fixing the meanings of professions and professionals. In this respect, as in others, exclusion of women genders professions and professionals. Take for example the case of the curriculum document for class 2 (GoP 2002c: 12–20). One of the specific aims in the context of creating social awareness is awareness of different professions. The list of professions (some of which are actually trades, though the text does not make this distinction) includes machine operator, mechanic, air hostess, driver, and bank employee. Section 2.3 of the same document recommends introduction of people in other professions especially agricultural worker (*kisan*: male), shopkeeper, barber, woodworker, clothes washer (*dhobi*: male and not *dhoban*: female), iron-smith, builder, potter, doctor, teacher (*ustad*: male, not *ustani*: female) and watchman. Notwithstanding the fact the persons mentioned in 2.3 have no relevance to the professions mentioned in 2.2, what is notable is that the single "feminine" profession mentioned in 2.2 is also excluded. The professions that class 3 children are to be made aware of, according to the same document, are the woodcutter (*lakarhara*) and the grasscutter (*ghasiara*).

Furthermore, the document recommends that children be made aware of how their district is run. It recommends that they be made aware of figures such as the *nazim* (a district manager), *naib nazim* (deputy district manager), *tehsildar* (a nonofficer rank bureaucrat), *taluqadar* (nonofficer rank land clerk), police officer, and so forth. In each of these cases the person is a male.

The corresponding textbook for class 3 shows a faithful textual following of the guidelines given by the CD. In *Meri Kitab* (my book), the Urdu reader for class 3, there is only one essay in which a woman is shown as a lead character. Similarly, there is only one essay in which the mother—and only to teach her kids household wisdom—plays the role of the knowledge provider. There is one poem with a lead female subject, and in this too the girl is confined to the four walls of the home applying henna while the brother runs around in the market. In the pictorial story in the reader (98), women are shown cotton picking (not one of the professions listed for them by the CD) while men are shown as industrial workers, tailors, and so forth. Eventually, the book introduces a female nationalist figure, Fatima Jinnah, sister of Muhammad Ali Jinnah. The text goes on to tell the students that "initially she used to serve her brother (*khidmat karti thien*) at home. Then she used to travel with the Quaid i Azam. She also used to help him in national duties. And she used to take care of his every need" (Punjab Textbook Board n.d.c.: 124). She is similarly articulated in the Urdu textbook for class 3 under the new textbook reform (PTB n.d.c: 59, 2009a: 85). A similar pattern of articulatory representations can be noticed in the new textbooks produced under the 2004 reform.

Finally, even with the overwhelming presence of religious and nationalist leaders in the textbook, women are by and large excluded from this category of subjects too. The significance of what it means to be a national or religious leader is thus fixed through exclusion and classification. First, in this respect, the students only get to know about women who were either prominent in a nationalist sense or in a religious sense. Here too the dominant articulation is in line with the stereotypes and chauvinism mentioned above. Two of the women who are most present in the texts are Hazrat Khadija with the title of *Ummatul Momineen* (mother of the faithful) and Fatima Jinnah with the title of *Madre Millat* (mother of the nation). Thus the ideals held up for young female students to follow are two mothers, one of Muslims and the other of the nation. The same level of reverence shown to these women once again makes the time boundaries fluid, as discussed above.

Other women mentioned in the text are either from the Prophet's (PBH) family (Ayisha, Zainab, etc.) or were active in the nationalist movement that led to the creation of Pakistan. In both cases, the frequency of their appearance in the text is pitifully low. This is specially the case with the women in nationalist leadership roles. Apart from Fatima Jinnah, there is scattered mention of a few others such as Lady Noon, but only in passing. It is interesting to note that two women leaders during and after the nationalist movement who worked on behalf of the women of Pakistan both within and outside the legislature, namely, Begum Jahan Ara Shahnawaz and Begum Shaista Ikramullah, are excluded from the texts. The new curriculum and textbooks under the 2005 reform minimally included these women leaders in the texts (PTB 2009b: 54). During interviews, I asked students if they knew of any of these ladies, and not surprisingly the answer was always negative.

Let me give a brief overview of the Urdu textbooks from classes 1–8 to map the presence of women in leadership roles in these texts. In texts prescribed for classes 1–2, there are simply no women in leadership roles. The only two figures that make it into the text are Prophet Muhammad (PBH) and Jinnah. The class 3 textbook features two women, Hazrat Khadija and Fatima Jinnah. The class 4 textbook features one (Ume Saleem, a Madinite mother who offered her son as a warrior for the battles of Islam). The Urdu text for class 5 does not feature any female leader, religious or national. Hazrat Khadija and Fatima Jinnah make it once again into the class 6 text. Finally, Hazrat Ayisha[7] (wife of Prophet Muhammad [PBH]) and Hazrat Zainab (granddaughter of the Prophet [PBH]) are included in the class 7 and 8 Urdu textbooks respectively. A total of five female religious or national figures in eight textbooks are presented as role models for students in eight years of education. In each of the cases, students come to know of the female figures in Islamic and in Pakistan's history only as cohorts or appendages of male religious and nationalist leaders.

The revised textbooks under the 2005 reform show a marked improvement over the 2002 curriculum in that the new textbooks have a lower militaristic content, especially in terms of the graphic representations of military battles and personnel. However, with respect to the presence of female role models, textbooks under the new reform do not fare any better than the preceding ones. For example, in Urdu textbooks for classes 1–8, there are a total of two women presented in a leadership role, namely, Fatima Jinnah and Khaula Bint Azwar. The former was the sister of Muhammad Ali Jinnah, while the latter is articulated as a warrior. Urdu textbooks for classes 7 and 8 have one chapter each that discusses the role of women during the independence movement and in the battle of Karbala respectively. In a nutshell, the representation of women in leadership positions by the textbooks and curricula has gone down under the new reform. Even key female figures such as Hazrat Ayisha (wife of Prophet Muhammad [PBH]), and Hazrat Fatima (daughter of Prophet Muhammad [PBH]) do not find adequate mention in the textbooks (Urdu textbooks for classes 1–8, 2004 reform).

Chapters on Hazrat Khadija in various textbooks, for instance, open with a description of praise for her feminine attributes and qualities. She is neither described nor praised for her economic status or know how. In fact she is shown to be looking for an honest man who could run her business. In other words, she is depicted as an insecure woman in need of male assistance to help her cope. The chapters conclude with examples of her supreme sacrifice when she gives her entire wealth to her husband to be spent in the service of Islam. The female subjectivity constituted by the text through this portrayal is of a subject who needs male help and support and who would not hesitate to sacrifice her material belongings for the nation, country, or religion. Such a subject and subjectivity orders the discourse by way of hierarchization and normalizes the classification.

Similarly, Fatima Jinnah, given the title of *Madre Millat*, is always shown as an appendage of Jinnah. Though the texts at times mention that she was a dentist by education and profession, her professional identity is never discussed except in the context that she gave it up to serve her brother and her country. She, like Khadija, also sacrificed her own interests for the male leadership figure (by giving up her dental practice). The discursively constituted identities of these women are the role models that girl students are encouraged to follow. While the virtues of statesmanship, generalship, and entrepreneurial knowledge are articulated as male attributes, women are ascribed virtues of sacrifice, domesticity, nurturing, and motherhood. An example of such articulation is also found in the CD for Urdu for classes 4 and 5. In its guidelines for the topic of *wafa shuari* (faithfuness), the CD suggests inclusion of an essay with Fatima, daughter of Prophet Muhammad (PBH), as the central figure. However, for an essay on the topic of *sachi dosti* (true friendship, homo-sociality), the recommended central figure is Abu Bakr, the companion of the Prophet (PBH). In comparison to the female, the male presence in the text is much more pronounced. Similarly, a whole chapter in the Urdu textbook for class 6 under the 2004 reform is devoted to *khana dar* (housekeeping) in which the mother teaches her daughter how to be an ideal housekeeper (wife) (PTB 2009a, p. 39–40).

Apart from Prophet Muhammad (PBH), Mohammad Ali Jinnah, and Muhammad Iqbal who are included in all texts without exception and in many ways, men are portrayed in the text in both religious and national leadership positions.[8] The articulation of patriotism draws heavily on the nodes of nationalism and religion. This particular articulation works through an overwhelming inclusion in the texts of men in military leadership roles and tales of valiance, bravery, and the ultimate sacrifice. Starting from class 3, the martyrs for Pakistan who were awarded the *Nishan e Haider*[9] appear prominently in all Urdu textbooks. As mentioned earlier, the articulation of military heroes is gendered to the core. These military heroes, to begin with, are repeatedly referred to as the "sons of Pakistan." Other accolades include lion, eagle, lion hearted, valiant, young, brave, and fearless.

These heroes fight and lay down their lives to safeguard the honor of the motherland and the nation against the Hindus who are assigned female attributes such as cowardice and fear. The Indian soldiers are referred to as *makkar aur ayyaar dushman* (a "cunning and devious enemy"). Their overwhelming presence and gendered articulation in the text has two functions: first, their inclusion excludes women and all those who have womanly or feminine attributes from the definition of a true Pakistani; second, it also serves to normalize militancy and militarism wherein these tendencies seem absolutely natural, a way of life, and accepted without question by the majority of the subjects. In the following space, I will discuss how the meaning fixation of the discourse is normalized through classifications by the texts.

*Normalization*

In this section, I discuss various articulations that are normalized by the social studies and Urdu texts. However, before I discuss the range of signs that are normalized by the texts and the processes through which normalization is achieved, let me briefly reiterate what I mean by *normalization*. Normalization refers to the definition of "normal" by the text. In a poststructuralist sense, it means the establishment of measurements, hierarchies, and regulations around the idea of a distributionary statistical norm within a given population—the ideas of judgment based on what is normal (Ball 1990: 2; also see Covaleski 1993). In simple words, it represents all those meanings that are fixed through various means by the discourse and made to look as if they are natural, ever present, and given. An impression of a broad consensus is constructed about their meaning, and large segments of society do not question them.

These meanings then regulate and govern all relations within the society. Any other meaning or interpretation that falls outside this normality becomes an aberration or deviance. All those who are found to be deviant are (or have to be) disciplined. I must mention that just as the meanings fixed for various signs keep shifting in accordance with the shifts in nature and/or the influence of the discourses that fix the meaning, the nature and range of signs to be normalized also changes. In Pakistan's case, the measurement of nationalism, patriotism, and what it means to be a citizen is accomplished by normalizing militarism, authority, discipline, and a gendered social hierarchy. I will now discuss how, through normalization, gendered subjects are constituted by the educational discourse in Pakistan.

# Normalizing Militarism

Militarism is normalized by the texts in many ways. At times it is conveyed in subtle ways through poems and isolated verses, mainly by figures such as Iqbal and Hali whose nationalist credentials have already been established by the text. At other times, militarism is normalized through narratives of epic proportion from the period of early Islamic history known as the *Ghazvat*[10] (literally "battles," but the reference is always to the battles that were fought either during the lifetime of the Prophet (PBH) or in the years following his death). The battles and wars of early Muslim adventurers in India such as Muhammad bin Qasim, Mahmood Ghaznavi, Ahmad Shah Abdali, and Sultan Tughlaq are also used to this end. At yet other times militarism is normalized through stories from the three wars between India and Pakistan. The stories of the sacrifices by the valiant sons of Pakistan who were awarded the *Nishan e Haider* also work to this end.[11] Last but not least, essays, poems, and chapters on issues seemingly mundane and unrelated to war are peppered with articulations that normalize militarism and militancy.

The text seeks to normalize nationalism based on religion, especially as manifested in the two-nation theory discussed in detail earlier. It also seeks to normalize the Hindu-Muslim binary, the normalcy of war between India and Pakistan, the notion of jihad (through reference to Kashmir), and the normalcy of the state as the ultimate protector. The textbooks produced according to these guidelines not only normalize militarism but also gender it.

For example, in an Urdu textbook from 1970s, a gendered and militarist explanation of jihad is used to normalize militarism. In the handwritten comments and edits of the subject specialist (name withheld), the word *children* in the sentence "There cannot be any Muslim who has not heard of the word jihad and Muslims know very well the meaning of this word, but there may be some children who might not be clear about it" (PTB 1974: 12) is struck out and is replaced by the word *larkay* (boys). The addressee of the text is gendered and the whole lesson then proceeds to explore the militaristic dimension of jihad. Though war or military action in fact constitutes only one form of jihad, the text presents it as the only one, and it is boys who should understand this concept well.

It is interesting to see the metaphors used in the articulation of these heroes. First, as noted above, is the use of metaphors of animals of prey (lion, falcon, eagle, etc.). Then they are masculinized with metaphors of male sexuality. Mohammad Hussain Shaheed, for example, is described as having a sinewy body, big black eyes, and thick brows (PTB 2002d: 65). Then they are ascribed with Rambo-like superhuman qualities. In accounts of their martyrdom, each one of them fought with multiple bullets in him and fought until the mission was accomplished. Take, for instance, the account of Lance Naik Muhammad Mahfooz. As the narrative goes, "One of Mahfooz's legs was injured by a cannon round. But he kept on crawling on his chest and jumped into the enemy bunker. The enemy sprayed fire at him and a bullet hit his chest. But even in this condition he locked the neck of an enemy soldier in the 'vice' of his hands. Meanwhile another soldier fired on him and a third one drove his bayonet into his body. Even when he 'drank the cup of martyrdom,' the neck of the enemy was still in his grip. Even the enemy acknowledged his bravery" (PTB 2002: 83). In the new Urdu textbook for class 3, the story of Hawaldar Lalak Jan tells the students that "Lalak Jan was injured by a cannon ball but kept fighting. The company commander tried to stop him twice but the injured Lalak Jan kept up his assault on the enemy" (Meri Kitab for Class 3, n.d.: 98–99).

Finally, these martyrs are included in the list of revered people by adding the suffix of *rahmaatu'lla alaihi* ("May God have mercy upon him") which, according to Barbara Metcalf, usually follows "the names of saints, great religious authorities, and other deceased pious persons" (1990: xiv). The picture that emerges from this articulation is that of a masculine, brave, nationalist, and revered person. This is the true citizen of Pakistan as articulated by the discourse through school texts. What it excludes is a majority of subjects (especially women) who have no chance of fulfilling the criteria laid down by such an articulation. The masculine subject is positioned by the discourse at the epitome of nationalist and religious authority. The feminine subject is excluded from this realm and can never hope to make it into the list of either nationalist or revered heroes and thus is relegated to the list of second-class citizen.

## Normalization of Authority

Earlier I talked about how, through the engendering of professions listed in the textbooks, women are excluded from the articulation of a professional. Gendered articulation of professions and professionals also serves to normalize authority. Take, for instance, the same text that engenders the articulation of a district management position. The stated goal of the CD is to introduce the students to how their local government (district) functions. However, the way it is presented also serves to normalize authority. In the class 3 Urdu textbook (PTB 2002b: 98), for instance, pictures accompanying the essay on district management show three cops guarding the rather colonial office of the deputy commissioner. Pictures on the next page show male judges, a male lawyer, male convicts, and so on—this is incidentally also the longest essay in the book (about 5 pages: 109–14). The gender use in the essay indicates that the chief district officer (*zilaee nazim e aala*) is always a male, as are the district liaison officer, the police officer, magistrate, and district education officer (DEO). Even the deputy DEOs and doctors are males. The only female figure in the essay is a nurse. The message of the lesson is that authority lies with the district administration/bureaucracy/local state (which, incidentally, is actually the case) and that all authority figures are male.

The Urdu text for class 7 (PTB 2002f: 80) conflates the ideas of civic order and discipline with battle discipline by exemplifying the former with one of the *Ghazvat* during the Prophet's (PBH) time. The essentialized message is to respect authority and maintain a military-like discipline. An earlier class 5 Urdu text informs the student that though citizens are given rights by the state, those who "unduly" demand these rights are not "good citizens." The text also informs students that the attributes of a good citizen are obedience, discipline at all costs, national pride, love of community, national pride, and so on (PTB 2002d; cf. PTB 1974).

It is interesting to note the reproduction of the state in the above-mentioned text. As the father (the knowledge provider) explains the rights of a citizen to his son: "The first right of the citizen is freedom...second right of the citizen is that the state protects the honor, life and possessions of the citizen.... A citizen has the right to enjoy all those facilities that are available e.g. water, power, radio, TV, news" (PTB 2002d: 78). The passage, however, does not talk about justice as a basic right, nor does it explain what is meant by "freedom." In light of the above discussion, it can safely be presumed that the only notion of freedom that the "son" will take in is that of freedom from Hindus, the British, and so forth. Freedoms such as that of speech, assembly, or association are never mentioned. Nor does the essay talk about nondiscrimination on the basis of sex, color, creed, or religious beliefs as the basic rights of citizens. The text emasculates even the male subjects to constitute a docile subjectivity that can be disciplined and ordered. Authority appears to be normal and demanding of rights appears to be an aberration. What is also normalized is the authority of the state as one that safeguards and permits—both masculine attributes. The state in this articulation is not a provider or a guarantor of basic freedoms.

## Normalizing Power/Knowledge

It is also important to note that in all social studies and Urdu textbooks for classes 3–8 and 1–8, respectively, there is rarely a female figure that is cast in the role of a knowledge provider. In all but two cases the knowledge provider is a male figure—most often a father or teacher (in some cases a paternal grandfather or uncle). Knowledge, according to the text, is thus a male preserve (See for example PTB 2002d: 109–13, 2002e: 77–81, 2002f: 44–47, 2002g: 63). So far the authority figures constituted are all male: savior (the soldier/*mujahid*), administrator (bureaucrat), and knowledge provider (father/male teacher). A fourth authority figure that the text defines is the imam (the spiritual leader and one who leads the prayers in the mosque). The CD repeatedly states that the role and place of imam and *mouazzin* (the person who recites the call to prayer five times a day) is to be inculcated along with the relationship between home, school, and mosque (see for instance GoP 2002c: 3, GoP 2002d: 4). Spiritual knowledge and power are thus also gendered, and women pretty much excluded from this domain.

## Normalizing Gender Relations, Normalizing Women

Let me start with the articulation of two women featured in the Urdu texts to show how gender relations, subject positioning, and subjectivity of women take place in the texts. My first example is the narrative of Umme Saleem, the Madinite woman who was also a distant relative of the Prophet (PBH). The text, after establishing who Umme Saleem was and how she came to the fold of Islam, tells the young readers that after converting to Islam she did not remarry for a long time (until her children grew up). When she finally agreed to marry Zaid bin Sohail, a not very well-to-do Madinite, she chose to forego her *haq mehar* (the money that a man is obliged to pay to woman upon marriage) by saying "convert to Islam and I won't ask you for any *haq mehar*." The narrative goes on to tell the students that she was an extremely brave and patient woman; that she thought of and called her husband *majazi khuda* (God in this world). She not only trod the straight path but also guided her husband to do so. Finally the essay closes with praises for her: "She proselytized through her deeds. She was always happy to go for jihad where she used to provide water to the holy warriors, hand arrows to archers, dress the wounded and bury the martyrs. She inherited hospitality and domestic work was her second nature. Her life is still a beacon for us" (PTB 2002e: 134–36).

The second text is an excerpt from one of Deputy/Maulvi[12] Nazir Ahmad's works. This is the story of Asghari. In the same book that reified major Tufail as a lion-hearted *mujahid*, a wounded lion, and a falcon, we find the story of Asghari, entitled the "good-mannered daughter-in-law":

> The girl [Asghari], while she was at her mother's was like a rose in the garden or like an eye in the human body [an Urdu metaphor]. God had given her every skill in the book: wisdom, smartness, etiquette, a kind heart, sociality, fear of God, *haya* [a shy demeanor appropriate for women; kind of a womanly reserve]. Since adolescence she hated playfulness and pranks. She only studied or did domestic work—all women of the community loved her. [They used to say] lucky will be the house where she will go as a daughter in law. As she turned thirteen she was engaged to one Muhammad Kamil....Once married she very patiently and cleverly took over the charge of her in law's household. She used to spend frugally and saved some money on a daily basis from the household expense. From her savings she used to buy some household item every month. Soon the house had new things....everyone started loving Asghari. (PTB 2002j: 71–75)

These textual articulations are more or less representative of the way woman is articulated by the social studies and Urdu texts in Pakistan from class 1 to class 8. These texts also stipulate the subject positioning of women and gender relations. In other words, they fix the meaning of what it means to be a "good" Pakistani/Muslim woman or how women should be understood by both men and women. Any other meaning of woman is a deviation and is thus erased from the discursive domain or contested. Any woman who does not understand herself within these discursive confines is a deviant and thus likely to be disciplined through the subject relations.

It is abundantly clear from the preceding discussion that the educational discourse in Pakistan, conveyed especially through the Urdu and social studies curricula and textbooks, constitutes subjects in relation to each other and in relation to the state. It is also clear that the discourse orders the power relations by means of particular subject positioning and the constitution of particular subjectivities. An advantage of looking at the subject and subjectivity constitution from the poststructuralist perspective is that it reveals the multisexual and multigender nature of the subjects and the state. This rectifies the shortcomings of the traditional feminist analyses that, by focusing on single or bisexual subject categories, not only essentialize gender but also prove unable to explain why the state behaves in different fashions with respect to different genders.

What also emerges from the preceding discussion is that through totalization, classification, and normalization, the curricula and textbooks discursively constitute subjects that are positioned in the discourse in such a way that women, minorities, and other subaltern groups find themselves largely excluded from the definition and thus benefits of citizenship. Certain subjects (largely male) and groups (military, bureaucrats, religious scholars, etc., and men) are privileged over other subjects and groups. These few male subjects are mandated with tasks of discipline through surveillance. In a society like Pakistan, unique temporal and spatial surveillance techniques such as the panopticon of dress, *chadar* and *chardewari*, and the gaze are also discursively constituted, while some men and groups are vested with the authority to exercise this gaze to order society in such a way that it facilitates postcolonial governmentality (interests of peripheral capitalism, an authoritarian state, supremacy of religiopoly, and a militarized culture).

# Chapter 8

## Education and Gendered Citizenship in Pakistan

On March 11, 2004, the entire opposition in the National Assembly of Pakistan walked out in protest over the omission of Quranic verses pertaining to jihad from biology textbooks for classes XI–XII. According to the text of the news report:

> An opposition protest walkout from the National Assembly on Friday forced a government apology in what seems to be brewing controversy over how much jihad should be taught in the country's schools and colleges. All opposition parties, despite their own known differences over Islamization, joined the MMA-led walkout to protest against a parliamentary secretary's remarks during the question hour.... The inclusion of Quranic verses is not a requirement of curriculum," said a written reply from Education Minister Zubaida Jalal in reply to a question from Laiq Khan (MMA, Sindh) about whether and why Quranic verses had been omitted from biology books for the intermediate first year. However, in this case, the Sindh Textbook Board has shifted Quranic verses from the book of biology for classes XI-XII to the book of biology for classes IX-X," the minister said. While answering supplementary questions from MMA members, parliamentary secretary Jafar Hussain denied their charge that the government was omitting verses about jihad and Christians and Jews to meet what they called US conditions for helping the country's education sector. (*DAWN*, March 12, 2004, online edition)

As mentioned in chapter 4, according to the Human Rights Commission of Pakistan, 1,635 women were murdered in a five-year period from 2004 to 2009 in what was claimed to be honor killings. A majority of these women were murdered by their own fathers, brothers, husbands, sons, and relatives. Less than half of the accused have ever been held. In the same period, 3,669 cases of rape were reported; 1,898 were gang raped. Only 481 of the accused were ever held.

The reason why I mention the first incident is the anomalous nature of the act in the parliament of Pakistan. This might be the first time that Pakistani parliamentarians have ever staged a walkout on the issue of content of textbooks in Pakistan. However, what is more pertinent for me is the part of the news item that states that "all opposition parties, despite their own known differences over Islamization, joined the MMA-led walkout" (*DAWN*, March 12, 2004).[1] The questions that arises are: Why would the apparently secular political parties in Pakistan join the MMA to protest the omission of Quranic verses pertaining to jihad from the biology textbooks? Was it political expediency? Did they want to appease the *ulema* (religious scholars) of the MMA? Or have these parties suddenly started caring about the issue of jihad (especially in the textbooks) despite their secular manifestos and record?

As I have argued earlier, the episode in the parliament of Pakistan is clear evidence of the entrenchment of the religiopoly discourse in Pakistan. This incident also vindicates the argument that subjects and subjectivity (political in this case) are constituted by discourses and that the subject positioning within the discourse organizes the power relations between subjects and between subjects and the state in the form of governmentality. The stance of the MMA and other political parties in Pakistan on the issue of omission (and their insistence on inclusion) of references to jihad in textbooks clearly shows the importance of texts to the type of subject constitution desired by the discourse. As I argued earlier, a combination of militarism and patriotism defines a nationalist citizen-subject. Thus any effort to displace this meaning is vehemently resisted by other subjects in the discourse. Although it is not clear what part the female minister of education has played in the effort to resist or displace such meaning fixation, her agency in this respect can nevertheless be speculated upon, along with the agency of other subjects (women, minorities, etc.).

The Human Rights Commission of Pakistan data on honor killing indicates that it is on the rise in Pakistan. Honor killing as a phenomenon is certainly not new to Pakistan: it has been happening in the areas that now constitute Pakistan for a very long time. What is different is the sheer increase in the number and visibility of such killings and more importantly the level of ease with which it is now accepted by the society at large. While earlier such killings in the name of honor were fewer and restricted to the rural and tribal areas, it has now become a widespread phenomenon that is also present in the urban areas as well. In the context of the present argument, this increase in numbers and visibility and its overall acceptance not as an aberration but an acceptable norm is clear evidence of the exercise of power through discipline by subjects that are privileged by the discourse. The acceptance of this heinous act of barbarity shows how certain acts that serve the discipline and surveillance functions, no matter how abhorrent, have come to be normalized to the extent that they not only become a part of daily life but also that any effort to resist or displace them is met with stiff resistance, often in the name of religion, culture, or a perverted notion of honor and tradition.

Pakistan today has a quasi-democratic political system. It is one of the seven nuclear powers. It has one of the larger standing militaries in the world. It is an important player on the world scene and has been granted the status of a major non-NATO ally by the United States. Since the coming into power of the current regime in Pakistan (and since the 9–11 incident), the economic situation (current and forecast) has relatively improved. The present government misses no opportunity to tout the fact that the foreign exchange reserves have been replenished many times over since it took power. At the same time, it is a common fact that as many as 39 percent of Pakistanis are living below the poverty line, that the health and education sectors have gone from bad to worse, that ethnic and religious cleavages have deepened, and that obscurantism and militarism are on the rise.

The state of women of Pakistan is a cause for deep concern. Despite official figures and claims that women's enrollment at all levels of schooling has increased many-fold and that women now have equal opportunities in the educational, vocational, and professional realms, these advances have not resulted in the empowerment of women as is evidenced by the increasing numbers of dowry deaths (commonly known as "stove-explosion" deaths, where it is inadvertently the daughter-in-law who gets killed when the oil stove suddenly bursts), honor killings, acid burnings and slashing for dressing "inappropriately," discrimination in employment and educational opportunities, to name only a few. As I was told by a number of women who I interviewed, the social and economic sanctions for dressing inappropriately are not only limited to those who wear Western dresses or who do not cover their heads. Even those who wear *hijab* are also victims of discrimination on the basis of dress and/or religious beliefs. One woman told me of the hardships she faced in first getting a research job and then holding onto it as she chose to wear *hijab* because of her religious convictions.

At the same time, the past few years have seen a remarkable practice of agency by the women of Pakistan in the legal, juridical, economic, and political realms, through which they have contested the existing gender relations and also confronted the state to contest for equal citizenship rights. We also see the new phenomenon of the emergence of right-wing women onto the scene. These women endorse a set of beliefs firmly grounded in a particular interpretation of Islam where women are not equal to men. The overall picture is complex and perplexing. Also in the gambit of differential subject positioning in the discourse are the members, both male and female, of minorities. Foremost in this respect are the Qadianis, whose exclusion from government and other jobs is one of the major demands of the religious right in Pakistan, and other religious minorities like Hindus and Christians who are increasingly finding themselves excluded or marginalized. Here the male-female gender distinction blurs to indicate multiple genders constituted by the discourse. The male members of the minorities find themselves excluded from the meaning of "Pakistani" (itself fixed around the nodes of religion and nationalism) just as the women of Pakistan do. The complexity of the situation brings me back to the broad questions that I raised earlier in the study:

1. How are subjects and subjectivities, especially gendered subjectivities, constituted discursively in the educational discourse in Pakistan?
2. What are the nodal points of the discourse around which the meanings of much of signs are fixed by the discourse?
3. Which signs are privileged and which are not? Furthermore, which signs are pushed out into the field of discursivity and how do they form resistance to the discourse?
4. What discourses does the educational discourse in Pakistan draw upon to constitute subjects and subjectivities?
5. Does education in Pakistan empower women?

In the light of the discussion in the preceding chapters, let me answer these questions.

Educational discourse in Pakistan discursively constitutes various subjects and subjectivities in relation to each other. Particular subject positioning determines their relationship with the state. In this process, the educational discourse draws upon other discourses, for example, the legal, constitutional, media, economic, and political discourses intertextually, i.e., it borrows (and lends), builds upon, and contests various meanings that are fixed for various signs (words, concepts, notions, etc.) by other discourses. The meanings fixed by the educational discourse discursively and intertextually are in relation to various nodes and nodal points (central concepts such as nation, Islam, etc.). When the discourses (especially educational discourse, which is the focus of the present inquiry) assign meanings to signs, the signs become elements, i.e., those whose meanings have been fixed by the discourse. This exercise of meaning fixation is known as *articulation*.

There are however, some signs that either escape this meaning fixation exercise or fall outside the assigned meanings. These signs are pushed out to the field of discursivity. These signs resist meaning fixation and produce resistance to the whole exercise. This articulation produces the knowledge through which subjects come to know and understand their own self and that of the others. It must, however, be kept in mind that meaning fixation is an ongoing process and that the articulated meanings are never fixed or final. The meanings are contingent on the relations between subjects, historicity, and the shifts or transformations that the subjects bring about in the discourse itself. In other words, discourse constitutes subjects and the constituted subjects in turn bring about changes or transformations within the discourse that in turn begets further meaning fixation and articulation.

The main reason behind this exercise is to constitute docile bodies that can be ordered/organized with minimum force and optimal economy so that their biopower can be harnessed, exploited, and put to use. Once these docile bodies are constituted, they are ordered by means of discipline that in turn depends largely on surveillance. As Foucault tells us, in the modern world (in contrast to the monarchical era), discipline and surveillance do not depend on physical measures. Instead a generally invisible gaze, the ever-present invisible eye, performs the function of surveillance. Any deviance, resistance, anomaly, or aberration is detected even before it happens.

I argue that while Foucault's argument is conceptually sophisticated enough to explain the constitution and control of subjects in developed societies, there is a need to further develop it to explain similar processes in the postcolonial world. As I show in my examination of women and the state in Pakistan, the state often uses surveillance and physical violence simultaneously. Similarly, the Foucauldian notion of gaze also has to be developed further to answer questions about its applicability in postcolonial societies. The notion of gaze that is inherent in the Foucauldian conceptualization rests on the official or officially appointed gaze. Building upon the Foucauldian notion of surveillance through gaze, I argue that, in the case of Pakistan, the discourse constitutes subjects and subjectivities and positions them in such a way that the task of surveillance is carried out by almost all subjects in relation to each other. Ordinary members of the society are delegated the task of carrying out surveillance activities in relation to others. Whereas in the developed and industrialized world, surveillance is mostly done through unobtrusive mechanical means such as debit cards, hidden cameras everywhere, and records such as credit reports and surveys, surveillance in many developing societies such as Pakistan is done through the gaze of certain subjects that have been privileged by the discourse through subject positioning. For example, in Pakistan all men (in general) are privileged by the discourse to exercise surveillance over women. Similarly, all subjects (once again males or Muslim males) are privileged to exercise surveillance over religious minorities.

My examination of education discourse in Pakistan shows that subjects and subjectivities are constituted in and through the Curriculum Documents and textbooks (together referred to as "the text"). From a poststructuralist, feminist epistemic position, I argue that the subject constitution in and through texts is essentially gendered. The text assigns gendered meaning to various signs around the nodal points of Islam and Pakistani nationalism. Through these meanings, subjects come to understand their own selves and their relationship to other subjects and the state.

The way this meaning fixation/articulation is carried out works as follows: The discourse, in the first place, fixes the meaning of what it means to be a Pakistani—the citizen. The meaning of Pakistani is fixed (around the nodes of Islam in its various interpretations and nationalism) in terms of someone who is a nationalist, as one who has the capability to fight wars with the enemy, one who can contribute to the military, the economy, and the political and administrative/bureaucratic system of the state. The text then goes on to construct narratives based on the history of the two nodal points; Islam and nationalism. In a majority of these narratives, the person or subject who can perform the above-mentioned tasks is almost inevitably a male figure (or occasionally a female figure with male attributes). Thus the first meaning fixed for the sign Pakistani is *male*. Other meanings include *brave, patriotic, anti-India, anti-Jew, militaristic,* and above all, *Muslim.*

Inherent in this articulatory process are at least three other articulatory processes that are carried out simultaneously. First, in order to articulate the Pakistani citizen as described above, texts use binary construction to constitute an "other" in reference to whom the self could be defined. This other is constituted as someone diametrically opposite of the Pakistani constituted by the text. This other is inadvertently the infidel, aka the Quresh of Makkah, the Hindu Rajas of India; the Hindu politician, especially those belonging to the Indian National Congress in the pre-1947 period; and the Hindu politicians, military, and the general population after 1947. The examination of the social studies and Urdu texts shows that the meaning of this other is fixed in feminine terms as compared to the masculine self of the Pakistani citizen. Other attributes of the nemesis of the Pakistani self are coward, sinner, schemer, and so forth.

This gendered articulation of self and other not only reinforces the masculine articulation of the Pakistani citizen; it also assigns the feminine attributes of the other—i.e., Hindu, Jew, etc.—to a number of subjects. This is the implicit, hidden subtext of the discourse. All those who match or possess the attributes of the other are automatically and by default articulated as other. Because the meaning fixed for the nationalist and religious other are predominantly feminine, anyone with these attributes becomes the other domestically. The women of Pakistan are thus othered through this process. Furthermore, since one of the major attributes of the Pakistani citizen is being Muslim, minorities are also othered and thus excluded from the definition of the Pakistani citizen.

At another level, this articulation is reinforced by fixing meanings to what it means to be a woman in Pakistan. The meaning fixation of the sign "woman" is also done in relation to the nodal points of Islam and nationalism. My research shows that the texts draw upon narratives from early Islamic and prepartition Indian history to fix the meaning of Pakistani woman. Only those parts of the narrative that are in consonance with and that reinforce the earlier articulation of the Pakistani citizen become a part of the text.

Thus, narratives and examples of domesticity, docility, sacrifice, and subservience to males are drawn upon from these historical periods to signify what an ideal Pakistani woman means. My examination of the social studies and Urdu texts also shows that any women outside these assigned meanings become anomalous and thus an aberration, or in other words, the other. Thus, just as the ideal Pakistani citizen has to fight the nationalist and religious other, he also has to fight the gendered other. This gendered other is also the one on whom constant surveillance has to be done. She has to be in the constant gaze of the patriotic Pakistani citizen. Any deviation from the fixed meaning is to be considered deviance and thus liable to be corrected.

As against the industrialized world, in developing societies like Pakistan, physical violence that results from a consciousness fragmented by the colonial projects as well as the gendered articulation of nationalism is still very much a rule. The male subject acquires a fragmented and often violent consciousness as a result of bisexual subject positioning with respect to the state. He is the superior citizen delegated with the surveillance gaze and mandated with limited powers to discipline the gendered other. With respect to the state, he is positioned as feminine needing the protection of the state from the Hindu, the Jew: from the enemy within and outside. The subject constituted by the discourse thus has multiple gendered subjectivities.

Once articulated, the subjects are then positioned in the discourse hierarchically where those higher up in the hierarchy are privileged over those lower down in the hierarchy. A poststructuralist feminist reading of the social studies and Urdu texts in Pakistan shows that the social and national reality constructed by the these texts has the military and the state at the top, followed by selected male members of the society, with women and minorities at the bottom.

Militaristic, masculine, and nationalist articulation is then normalized by the educational texts in conjunction with the media, judicial, and constitutional texts. This normalization is carried out by means of overt as well as subtextual representation of masculinity and patriotism, coequating of religion and nationalism as one and the same, and a multilayered, gendered constitution of subjects where the feminine subjects is in need of protection from the masculine and militaristic state.

Thus, as we see in Pakistan, the military and militarism of all shades and hues—whether in the form of religious fanaticism, violence against women and children or minorities, support for *jihadi* organizations domestically and internationally—has come to be seen as normal. An example of the normalization of the military and militarism is the fact that Pakistan might be one of the few countries in the world where the military/defense budget is not under the purview of the parliament even when the country has a democratically elected, civilian government.

The social studies and Urdu texts in Pakistan construct a homogeneous and totalizing narrative of religion, nation, and nationalism grounded in the Orientalist epistemic framework that in its present form includes the exegesis of globalized capitalism. Intertextually drawing upon the rubric of the Orientalist discourse, the texts strive to achieve theoretical and explanatory homogeneity. In doing so, all differences are obfuscated and veneered with the gloss of oneness. Thus, not surprisingly, any narrative that does not fit or even hints at disrupting the homogeneity of theory or explanation is excluded from the grand narrative.

The social studies and Urdu texts construct a metanarrative of religion and nationalism that includes only the masculine, militaristic, and nationalist narratives from past and present. In this metanarrative, both brain and brawn is the forte of the male Muslim Pakistani citizen. The sources of knowledge are limited to nationalist and religious scriptures. Thus, excluded from the metanarrative are women, dissidents, and minorities. Alternative religious and nationalist texts are also excluded and condemned. Experience is decried as unauthentic and unscientific and thus not a credible source of knowledge. Furthermore, the texts graphically and discursively present knowledge construction and knowledge provision as essentially male activities.

Thus, not surprisingly, when passed on to the learners, this knowledge does not seek to empower the subjects but to order them. It is in this process of ordering and disciplining the subjects that some are selectively empowered (only in relation to other subjects). The educational discourse in Pakistan along with its disciplinary institutions (schools, colleges, universities) and surveillance techniques (the gaze, examinations, teachers, etc.) does not empower the subjects that it constitutes to attain a consciousness of the self; an understanding of their raison d'etre. It empowers few in relation to many so that the docile bodies can be economically ordered into being productive subjects.

The social studies and Urdu texts in Pakistan thus constructs subjects that are docile, who lack the ability to imagine, question, and analyze; who are prone to take instructions rather than innovating solutions; whose loves and hates are on demand and ordered from above; whose imagination is based on make-believe, grand narratives of historical grandeur and glory. They constitute subjects with a fragmented consciousness and often violent self.

At the same time, these texts by default also constitute resistance to the these subjectivities in form of female activists of the Women's

Action Forum, the Sindhiani Tehrik, and countless, nameless other women who in their own ways not only resist but have brought about little kinks in the discursive armors of the discourses that constitute them. The infamous *hudood* ordinance, for example, was overturned in 2006 after a long struggle waged by the valiant women of Pakistan. While political and social activists and academics have been at the forefront of such agency for change, female teachers and students find spaces for resistance in the very discourses that constitute their subjectivity in the first place. Though the educational opportunities afforded to them are scarce and unequal, these subjects express their agency in numerous ways, whether it be better academic performance as compared to their male counterparts or in devising innovative teaching and learning strategies that challenge the docile subjectivity constitution and unfavorable subject positioning. Through such agency, challenges are mounted to exclusionary practices and strategies and to the unequal citizenship status accorded to women. Together these women have contested the discursive meaning making that makes them invisible. They have resisted and redirected charges of being "Westernized," antisocial, and at times anti-Islam to create unique and innovative channels of resistance and agency.

In them the future of the country lies.

# Notes

## 1. Contextualizing Articulations of Women in Pakistan

1. Fixation of meaning for a particular sign (word).
2. Scholarship in this respect ranges from theorizing the relationship between women and the state (Mumtaz and Shaheed 1987; Khattak 1994; Zafar 1988, 1991; Jalal 1991; Rousse 1994; Hussain, Mumtaz, and Saigol 1997; Jamal 2002), to women, religion, and law—especially the Sharia (Islamic law)—(Khan 1992; Gardezi 1994; Jahangir and Jillani 1990; Rousse 1994; Saigol 1994; Shaheed 1998; Weiss 1986, 1998; Zia 1994; Haq 1996; Saeed and Khan 2000), women and media (Hussain and Shah 1991), agency (Shaheed 1998; Sathar 1996), fertility (Sathar 1996), women, diaspora, and postcoloniality (Jamal 1995, 2002), issues related to identity (Khan, Saigol, and Zia 1994; Naseem 2004, 2006), and education (Saigol 1993, 1995; Naseem 2004, 2006).
3. For examples of early feminist writing see Mumtaz Shahnawaz's *Heart Divided* (1957/1990) and J.A. Shahnawaz's *Father and Daughter* (1971).

## 2. Michel Foucault and I: Applying Poststructuralism to the Constitution of Gendered Subjects in Pakistan's Educational Discourse

1. A growing number of studies have engaged with poststructuralist theories of all shades and hues to examine and explore educational questions and problems. Poststructuralist concepts have been used to analyze power dynamics in inner-city schools in Brazil (DaSilva 1988), teacher's education (Popkewitz 1987), examination and testing (Marshall 1990), educational research (Marshall 1990; Lather 1991; St. Pierre and Pillow 2000), discipline and power in education (Covaleski 1993; McDonough 1993), curriculum (Apple 1990, 1991, 1995; Beyer and Apple 1988; Cherryholmes 1987, 1988; Bjerg and Silberbrandt 1980; Popkewitz 1987; Freedman and Popkewitz 1988; Curtis

1988; Blades 1997), and student identity (Mayo 1997). Although contributors to Peter's (1997) volume on Education and the Post-Modern Condition engage and converse more with Lyotard's conceptualization, a number of them engage with Foucault in more ways than one (see Bain, Nicholson, McLaren, Marshall in Peters 1997).

2. By *normalization* Foucault means the establishment of measurements, hierarchy, and regulations around the idea of a distributionary statistical norm within a given population—the ideas of judgment based on what is normal (Ball 1990: 2. Also see Covaleski 1993).

3. *Discipline* in this context is used as a verb (not noun) as in disciplining in schools

4. It refers to the governmentalization of the administrative state of the fifteenth and sixteenth centuries.

5. Foucault has been criticized for being gender neutral (at best) and gender blind (at worst) by feminists. See Sawicki 1988; Bartky 1988; Fraser and Nicholson 1990; Naseem 2004.

6. *Femocrats* is the term used for feminist women bureaucrats who work to bring about a change from within the governmental institutional structures. The term was made popular after a number of Australian women bureaucrats were successful in getting concessions for women. See for instance Franzway 1986; Watson 1990; Pringle and Watson 1992.

7. Not all femocrats are poststructuralists.

# 3. The Education System and Educational Policy Discourse in Pakistan

1. Biopower is a new type of power that emerged in the nineteenth century and that was necessary for the development of capitalism. It exerts itself on the organizational and disciplinary aspects of life. Its basic aim is to subjugate and control and thus create docile bodies. According to Foucault (1990b: 140) biopower refers to "the controlled insertion of bodies in the machinery of production and the adjustment of the phenomena of population to economic processes."

2. Jointly administered by both the federal and provincial governments.

3. Prior to 1971, East Pakistan was the fifth province.

4. Data Source: The survey's estimates are based on the data of 36,272 sample households enumerated from July 2007 to June 2008. Findings are presented in the form of proportions and percentages to provide for all-purpose employability. The population of Pakistan as per the Planning & Development Division's projection is estimated at 160.97 million on January 1, 2008. The same statistic has been used in arriving at absolute numbers in the report. (Source: Government of Pakistan, Statistic Division (http://www.statpak.gov. pk/depts/fbs/publications/lfs2007_08/results.pdf).

5. Neither the Ministry of Education, Government of Pakistan (http://www. statpak.gov.pk/depts/az/az.html), nor the Economic Survey of Pakistan (2002) give the definition of literacy to explain their reported or projected estimates.

6. With the exception of the (Zulfiqar Ali) Bhutto period (1972–1977), educational institutions in Pakistan have existed in both domains.

7. Jinnah's address to the conference had anticolonial overtones, yet the undertones were conditioned by the nationalist/modernizing discourse. Speaking to the conference, he said, "You know that the importance of education and the right type of education cannot be over-emphasised. Under foreign rule for over a century, in the very nature of things, I regret, sufficient attention has not been paid to the education of our people, and if we are to make any real, speedy and substantial progress, we must earnestly tackle this question and bring our educational policy and programme on the lines suited to the genius of our people, consonant with our history and culture, and having regard to the modern conditions, and vast developments that have taken place all over the world. There is no doubt that the future of our state will and must greatly depend upon the type of education and the way in which we bring up our children as the future servants of Pakistan. Education does not merely mean academic education and even that appears to be a very poor type. What we have to do is to mobilise our people and build up the character of our future generations. There is immediate and urgent need for training our people in the scientific and technical education in order to build up our future economic life, and we should see that our people undertake scientific commerce, trade and particularly, well-planned industries. But do not forget that we have to compete with the world which is moving very fast in this direction. Also I must emphasise that greater attention should be paid to technical and vocational education."

8. I have dealt with this issue in detail elsewhere (see Naseem 1995).

9. An interesting example of this is the way selective verses from Quran, selective Hadith of Prophet Muhammad (PBH), quotations from Jinnah's speeches, verses from Iqbal's poetry, and speeches of the incumbent in office were used to construct docility and obedience as a religious and thus nationalist virtue. Academic texts were supplemented by the media texts. During Zia's reign, for example, just minutes before the national news bulletin on prime-time TV (and there was only one channel in those days) and radio, sayings of Prophet Muhammad (PBH), quotations from Jinnah's speeches, Iqbal's verses, and excerpts from Zia's own speeches were flashed in this very order. Interestingly, all of these emphasized obedience to the ruler of the day. Other than the overt inculcation of obedience as a virtue, the broadcasting of these messages also created a hierarchy in the minds of the viewers in which Zia as the ruler of the day was fourth in line (fifth if God is to be the first) after Muhammad, Jinnah, and Iqbal in terms of reverence. It was also during this time when the religious and nationalist discourses fused symbiotically that the words *Rahmatu'lla Alaihi* ("God may have mercy upon him," usually used for great religious authorities and saints) were added to the names of Jinnah and Iqbal, thus equating nationalism with religion.

10. Although most scholars read this as Zulfiqar Ali Bhutto's nationalist/populist/anti-Western rhetoric, there is more to it in my opinion. Taken into consideration in conjunction with the masculinist symbolism that was rampant in senior Bhutto's self-characterization and the titles that he bestowed upon

himself (i.e., *Sher e Pakistan*, "lion of Pakistan"), his cabinet colleagues (he publicly used to call Hafiz Pirzada *Sohna Munda*, "pretty boy"), and his opponent (words not suitable for an academic discussion) and his playboy lifestyle, it clearly shows that it was not only his nationalist/populist rhetoric that mandated *shalwar kamiz* as the nationalist dress. For an excellent exposition of Bhutto's self-characterization see Syed 1975. For an equally good exposition of his life style see Taseer 1980.

11. The quality and level of education imparted to the girls in the *mohallah* schools was abysmally low. Most often the teachers in these schools themselves had only high school education and no teacher training.

12. The policy details the following steps in this respect:

    "Introduced English, Economics, Pakistan Studies and Maths at secondary level in 140 *Madaris*. English, Economics and Computer Science at Higher Secondary level are being offered in 200 *Madaris*. In compliance with the present policy 8000 *Madaris* are to be facilitated to teach formal subjects from primary to higher secondary levels. The policy aims: To facilitate integration of *Deeni Madaris* with the formal education system. To teach formal subjects in 8000 *Deeni* Madaris to bridge the gap between Madrassah Education and Formal Education System. To open the lines of communication with *Ulama* who run the *Madaris* to impart formal education in addition to religious education for spreading of Islamic values at national and international level. To improve and update knowledge of their teachers in formal subjects through workshops at different parts of the country." (Ministry of Education, online)

13. In the last elections (October 2002) when the Musharraf government decreed that only those with a bachelor's degree can contest the elections, a number of politicians were debarred from candidature. However, all candidates of religious parties stood eligible on the basis of degrees issued by their respective *madrassahs*. It was also reported in the national press that a number of politicians also obtained degrees from *madrassahs* in exchange for hefty sums of money.

14. I remember that when I took the interview for admission to the masters program at the prestigious Quaid I Azam University, the grapevine was that unless we knew certain verses of the Quran by heart, we would not be admitted to the program as the dean and faculty of Social Sciences at that time thought it was imperative to admit only "good" Muslims. I also remember an incident where a senior professor at one of the departments told our female classmates that unless they covered their heads, there was no chance for them to secure an A grade, no matter how good their exam papers were.

15. A district is an administrative unit within the province.

# 4. Women and the State in Pakistan: A History of the Present

1. These include but are not limited to: governance and governability (Nasr 1992), state and democracy (Jalal 1995; Noman 1990; Shafqat 2002),

military and politics (Choudhury 1988; Dewey 1991; Jalal 1995; Rizvi 2000; Waseem 1987), nationalism and national integration (Jahan 1972; Khory 1997), constitutional dynamics (Azfar 1991; Choudhury 1963, 1969), political economy (Jalal 1990; Noman 1990; Sayeed 1980), Islam/religion/ Islamization (Esposito 1996; Ewing 1988; Jahangir and Jillani 2003; Kurin 1985; Richter 1986; Shah 1996), ethnicity (Amin 1988; Kazi 1987), and political parties (Afzal 1976).

2. Marxist (Alavi 1972, 1983, 1990); historical-structuralist (Sayeed 1967, 1980; Waseem 1994), revisionist historiography (Jalal 1990a, 1995), civil society (Malik 1997; Pasha 1997); institutionalist (Jahan 1972; Shafqat 2002).

3. Scholarship in this respect ranges from theorizing the relationship between women and the state (Gardezi 1994; Hussain, Mumtaz and Saigol 1997; Jalal 1991; Jamal 2002; Khattak 1994; Mumtaz and Shaheed 1987; Rousse 1994; Shaheed 1990; Zafar 1988, 1991); to women, religion and law, especially the *Sharia* (Islamic) law (Ewing 1988; Esposito 1996, 1998; Gardezi 1994; Jahangir and Jillani 2003; Khan 1992; Rousse 1994; Saigol 1994; Shaheed 1998; Weiss 1986, 1998); women and media (Hussain and Shah 1991; Zia 1994); agency (Kazi and Sathar 1991; Mumtaz and Shaheed 1987, 1991); fertility (Kazi and Sathar 1991); women, diaspora, and postcoloniality (Jamal 1995, 2002); issues related to identity (Khan, Saigol, and Zia 1994; Naseem 2004); and education (Naseem 2004, 2006; Saigol 1993, 1994, 1995).

4. In this section I retain the hyphen in post-colonial as Alavi used the term with the hyphen.

5. Broadly, the postfoundational position subsumes postcolonial, poststructuralist, and post-Orientalist positions (among others). Scholars working from this tradition take insights from Edward Said's Orientalism, Foucauldian and Derridian Post-structuralism and subaltern studies (Guha and Spivak 1988) to argue that in order to have a meaningful understanding of the postcolonial societies, the focus should shift from the Eurocentric notions of state. Furthermore, they argue that the colonial phase in the life of these societies should not be seen as a phase in progression from premodern to modern. Instead, it should be analyzed as an intervention that through a cultural project re-created Europe and its other and through which both are understood even today.

6. Said (1979) defines Orientalism as a style of thought based upon an ontological and epistemological distinction made between the Orient and the Occident. It constructs the Orient as the "other" of the Occident and presents it as lacking everything that the Occident has, e.g., civility, culture, rule of law, knowledge, and so forth. Orientalism has been the style of thought that has been used by both the colonialist as well as the colonized to study and understand the people, culture, and mind of the Orient. After World War II, modernization theory took up the reins of Orientalism from where the colonial project left off. In course of my research on Pakistan, *Orientalism* refers to the legacy of the colonial cultural project and the exigencies of the modernization discourse in the post–World War II era.

7. For instance, as Gyan Prakash (1994) argues in the case of India, while the postindependence nationalist historiography in India accomplished much

in terms of dismantling the exotic image of India constructed by the colonial literary project, it did so largely within the confines of the categories of thoughts and procedures of Orientalism (Prakash 1992).

8. Exceptions to this scholarship are Mumtaz (1999), Rousse (1999), and Saigol (1995).

9. I am particularly referring to the feminist scholarship here as different from feminist activism that has addressed issues related to minorities and has advocated for their inclusion in the national discourse.

10. While for some it represented the culmination of the independence movement into an Islamic state for the Muslim population of India (Choudhury 1963, 1969), for others it was a compromise by Jinnah who was more interested in getting an equitable share of political power for Muslims within a combined India (Jalal 1985). Yet for others it represented a schizophrenic personality caught between religious ideology and the demands of a modern nation state (Noman 1990). Those working from a historical structuralist perspective saw the new entity essentially as a modernizing force that was both an agent as had agency (Waseem 1994). For Marxist scholars such as Hamza Alavi (1971, 1990), the new state was a conduit for extraction and exploitation for peripheral capitalism.

11. Although, it was Raja Ram Mohan Roy at the forefront of struggle for the eradication of *sati* in India.

12. Mumtaz and Shaheed show that in India it was "the men of the supposedly backward colony, rather than the colonialists of the supposedly advanced countries, who supported the cause of women's franchise" (1987: 42).

13. In abolishing the practice of *sati*, the colonizers created the gendered "other" of the British male and female. The British civilized male vs. the native brutal male, and the Victorian "liberated" female as against the subjugated female. In the colonial accounts of *sati*, one finds no mention of the parallel British/European traditions of spinster aunt or widowhood linked to private property. For an excellent exposé, see Jack Goody's comparative analysis of domestic domain across three continents, Africa, Europe, and Asia (1976) Also see Goody 1983.

14. The colonial land administration system was enacted in the form of land distribution through the Permanent Settlement System, the *Ryotwari* System, and the *Mahalwari* System. For an analysis of these land administration systems, see Waseem 1994: 30–36.

15. While I agree with Mumtaz and Shaheed's analysis, I do not agree with their use of the notion of feudalism without qualification. I believe that in India, feudalism as a concept, like that of class, was a British import. The rural landowning structure of India was not analogous to that of the British feudal classes. This is not to say that it was not regressive with respect to women. The point is that it has been largely unexplored in terms other than the colonial/Orientalist.

16. Elsewhere Chatterjee has argued that it was the nationalist leadership that denied the colonial state the right to intervene in the realm of personal law. See Chatterjee 1994.

17. Mumtaz and Shaheed (1987), apart from the legal and economic changes, also attribute this to the loss of power of the Muslim males following 1857 war of independence. Also see Metcalf 1990.

18. The private conversations that I was exposed to while growing up in Pakistan were narratives of women who cooked (voluntarily and happily) for the large political gatherings of men. Or they were stories of women who entered the discourse of nationalism and the nationalist movements while men fought and gained concessions for the women from the colonial government.

19. The act was subsequently passed, but Jinnah, as the governor general, withheld his consent. Even once it was finally placed on the statue books, it was to supplement and not override the customary laws. In effect, it was discretionary for those families who wanted to give inheritance to their daughters (Jalal 1990: 87).

# 5. Subject Positioning and Subjectivity Constitution in Pakistan

1. For details and analyses of constitutional issues in Pakistan see Waseem (1994). Also see Azfar (1991) and Sayeed (1967, 1980).

2. The period of 20 years expired in 1993, and despite the fact that four different civilian governments have been in power, the reserved seats have not been revived.

3. Industrialists and *mandi* merchants were the main financiers of the PNA movement.

4. Zulfiqar Ali Bhutto's measures such as banning of alcohol and gambling and declaring Friday as the weekly holiday instead of Sunday have often been attributed to the pressure by the religious parties by most scholars. In my opinion, Bhutto could also sense the insecurity and anger of the people because of the above-mentioned incidents and thus resorted to symbolic Islamic actions.

5. *Hudood* is the plural of *hadd*, which in Islamic jurisprudence is "crime against the boundaries set by God," thus crime against Allah.

6. The *qanun e shahdat* (law of evidence) does not fall under the rubric of *hudood* laws.

7. Interest was renamed as profit. In principle, the investments/deposits made by the investors could either incur a profit or a loss, but in practice the banks fixed a floating interest rate that was always profitable for the investor. Thus interest was made kosher by calling it profit.

8. Rousse (1999: 61), on the basis of data collected by various human rights groups, reports that by the mid-1990s, 75 to 80 percent of all women in jails were held on charges of *hudood* offenses.

9. Under the old law, the complainant was under no obligation to produce four witnesses. Any evidence that could prove that the crime had been committed was enough to prove the case.

10. Mumtaz and Shaheed write: "Almost all religious groups and political parties welcomed the law. This was only to be expected. What is surprising is that even lawyers, particularly women lawyers, failed to examine the law. Perhaps the finality of the Martial Law Order, combined with the prevailing atmosphere of political despondency, accounts for all the progressive forces having

ignored" (1987: 101). In a similar vein, Jahangir and Jillani (2003) note that even after the death of despotic General Zia, and the overwhelming denouncement of the *hudood* ordinances by the Benazir Bhutto regime, intellectuals, jurists etc., the ordinances could not be repealed. Jahangir and Jillani attribute this to the presence of Ziaist elements in the parliament (and outside).

11. "Preparation for *zina*" (fornication) could be anything from sitting with men to a smile or anything that the men or the court deemed as inviting.

12. Literally, "veil." Most of the writers use the term *chador*, which denotes the Persian pronunciation. In India and Pakistan, the word is pronounced as *chadar* or *chaddar*. This is the form that I will use.

13. Literally, "four walls of the home."

14. Here I will not go into the details of the debate about religious and cultural articulation of *hijab* or *chadar*. Excellent material already exists on the debate. See for instance Fatima Mernissi (1987) for an excellent account of theological aspects of the debate. See Weiss (1986, 1992) and Shaheed (1991) for the discussion in the Pakistani context.

15. In 1980, directives were sent to the television authorities that all women appearing on television were to wear *chadar*. Services of those who refused were terminated. Similar directives were sent to all schools, colleges and universities, and government offices. Teachers were to set an example for the students by wearing *dupatta* (Mumtaz and Shaheed 1987: 77–78). What is ironic is that even the strictest interpretations of Islam do not require women to wear *hijab* unless they are in the company of nonfamily males (*mehram*). In Pakistan, all public schools are already segregated and with the exception of perhaps the old *chowkidar* (the guard) who watches the gate, no men are present in the schools or colleges. No directive even mentioned the Islamic *hijab* for men, i.e., lowering one's gaze in the company of women outside the family.

16. Here I have used the translation by Hafeez (1981: 60). Earlier translation is mine.

17. All these terms were used for members of Women's Action Forum (WAF), the most activist women's group during the martial law of Zia ul Haq.

18. While plays in this genre did not always project the feudal culture in a positive way, generally they portrayed this culture as a way of life.

19. Mumtaz and Shaheed note, "Television programs depicted women as the root cause of corruption, as those who forced poor men to take bribes, smuggling or pilfering funds, all in order to satisfy the insatiable female desire for clothes and jewellery." (1987: 82)

20. He was also a member of Zia ul Haq's hand-picked Majlis e Shoora, the consultative assembly.

21. Veena Hayat was gang raped by a group of men led by the son-in-law of the then president of Pakistan, Ghulam Ishaq Khan. The perpetrators were never apprehended, and a large section of the press and public laid the onus of the incident on Veena Hayat on the grounds that, being a socialite, her behavior invited such an act. Such justifications have become common in Pakistan ever since. The implication is that if a woman steps outside the boundaries drawn for her, then any act to control her is justified. But she did not step outside the boundaries: she was raped.

22. There is a wide-ranging view in Pakistan that the liberal press supported the Women's Action Forum (WAF) in its efforts against the despotic government of General Zia. To me this is a superficial observation. A deeper examination of the news media shows that what is usually taken as support was itself very gender biased (See for instance Hussain and Shah 1991). Also see Zia (1996).

23. In the Zainab Noor case, a *mullah* burnt the sexual organs of his wife by inserting an electric rod in her genitals.

24. In this case, Al Huda is different from Dr. Israr Ahmad's television program mentioned above. Here it refers to the network of schools that impart a particular brand of religious knowledge, usually to children (especially girls) of upper-middle and upper strata of the Pakistani society. These institutions are funded by transnational and diasporic finances and often feature speakers and teachers who live abroad.

# 6. Educational Discourse and the Constitution of Gendered Subjectivities in Pakistan

1. Here the notions of subject and subjectivity are used in a poststructuralist sense broadly meaning people and their identities. These should not be confused with disciplines or educational subjects such as history, geography, etc.

2. I have first-hand experience of this. At the time when India tested its nuclear devices for the second time and Pakistan was weighing its response, along with a handful of others I proactively argued that going nuclear was not a wise choice for Pakistan. We were at once dubbed as antistate, anti-Islam, Indian and Jewish agents; were titled as cowards, doves, etc. and were attributed with these feminine qualities. Wearing my hair in a ponytail almost immediately became a sign for weakness and feminine qualities.

3. Note in this respect the appellations of indigenously produced weapons, especially the missiles in Pakistan. These are named after Muslim warriors such as Shahabuddin Ghauri (the Ghauri range of missiles) or the weaponry used by Muslim warriors (e.g., Anza, the shoulder-held missile, is named after the lance of Ali, the fourth of the Pious Caliphs). Another missile series is named Shaheen (I&II). The symbol here is the eagle in Iqbal's nationalist poetry.

4. Aziz (1993) through a meticulous analysis of a large number of textbooks has shown the fallacy in such presentation in the textbooks. After listing every book that makes such or related claims, Aziz argues: "(1) It was not Pakistan Resolution but the Lahore Resolution....(4) It was not passed on 23 March but on 24 March. (5) It did not demand *an* independent state; the word 'State' was used in plural" (Aziz 1993: 146). Aziz writes further, "The resolution is so clumsily drafted that in the opinion of some careful scholars it is debatable whether it demanded independent status or suggested some kind of a confederation between the Indian state and the Muslim 'States' " (1993: 146). Jalal (1985) also takes a similar stance with respect to Jinnah's real intentions.

5. There is also an evident intertextuality between the literary and the educational discourses. This particular text for instance draws upon Mohammad

Iqbal's nationalist poetry especially his poem "*ho halqa e yarn to resham ki tarah narm/ bazm e haq o batil ho to faulad hai momin*" (In the company of friends he is soft like silk/ in a contest between right and wrong a true believer (Momin) is like steel). (Translation mine)

6. Urdu is by no means one of the most common languages in the world. The fact that Urdu broadcasts are listened to by immigrant Pakistanis in Asian, African, and European countries does not make it popular or common in these areas. Urdu was not the language of Bengali Muslims who were as much a part of the independence movement; nor has there ever been a consensus in Pakistan on Urdu as the lingua franca. The language riots in 1950s and 1960s were precisely over the issue of nonacceptance of Urdu as a national language; one of the major demands of Bengali nationalists in former East Pakistan was declaration of Bengali as an official language besides Urdu.

7. History was to show in 1971 that the separation of East Pakistan was based on the very differences Jinnah sought to minimize.

8. I borrow the notion of imagined and not imaginary boundaries from Anderson (1986). According to Anderson, imagined boundaries are real (unlike the imaginary boundaries) and are based on specific attributes that are imagined by a certain community that it intends to achieve.

9. Both generals claimed their coup d'etat as revolutions. Both gave debauchery of politicians as the raison d' etre for their intervention into politics. For Ayub Khan, see his *Friends not Masters* (1966). For Zia ul Haq see Aziz (1993). Both claimed to have saved the honor of the nations from the Raja Dahirs of politics.

10. In the postconquest time, Qasim, the class 6 social studies text tells us, was loved by the Hindus, who were sick of their Hindu rajas who treated them like animals. According to the text (PTB 2002e: 100; PTB n.d.h: 94–97), "The local population began to love Muhammad Bin Qasim....he was kind both to Buddhist monks and Hindu Barahmans. No harm was done to their places of worship rather he made grants to their temples...improved the condition of the labourers and farmers.... The Hindus began to look upon Muhammad Bin Qasim as their saviour." Even in cases where there is strong recorded evidence that Muslims were not gentle and tolerant of the beliefs of Hindus, they are not portrayed negatively. As Ali's analysis of textbooks shows, "Mahmud Ghaznavi's treatment of the priests and worshippers, and lack of respect for their beliefs, is not presented in negative terms as would have been the case had a Hindu ruler desecrated Muslim sacred space" (1993: 232) as was the case after the Babari Mosque demolition in December 1992.

# 7. Classification, Normalization, and the Construction of the "Other" in Pakistan's Educational Discourse

1. The commonly used phrase in the vernacular to connote "enemy" is *Yahood-o-Hanood*, literally "Jews and Hindus." As is the case with the collation of pre-Islamic Arabs and pre-Islamic Hindus in terms of collating

the boundaries of time in the postindependence period, it is the collation of Jews and Hindus that threatens the motherland. A recent example of this is the statements of Pakistani leaders expressing concern over the defense deal between India and Israel. This deal is being seen and articulated as a long-standing desire of Jews and Hindus to obliterate Pakistan. Pakistan and Israel do not have any bilateral dispute. The reasons for Israel's alleged desire to destroy Pakistan (through India) are not very clear.

2. Anthropological research in South Asia has shown that the menstruating woman is still considered unclean. The Islamic moral code forbids men to have sex with menstruating women. Menstruating women are also freed of the obligation to pray or fast during their menstrual cycle. Although feminist interpretation suggests that the purpose of these injunctions is to give women a break, the common perception among Muslim males is that it is because of the bad blood that the women discharge.

3. The text also omits to mention the services of the Parsi and Memon communities to the cause of the independence movement. It is a documented fact that these were the first communities who were asked by Jinnah to move their successful business from Bombay to Karachi even before Pakistan came into being (for details see the analysis by Venkataramani (1986) based on declassified state department documents).

4. This knowledge corroborates the argument made earlier in the chapter about the gendered articulation of Muslim warriors. In almost all texts, whenever the Muslim warriors were victorious, it was because of their inherently male qualities such as bravery, virtue, justice, generalship, etc. When they were defeated, it is always due to conspiracy, treachery, and betrayal and thus not their own fault. With respect to the "other," i.e., the Hindus, this Samson and Delilah articulation is reversed. Whenever the Hindus are defeated, it is because of their weakness. When they are successful, it is always because of unholy alliances, trickery, treachery, deceit, and backroom tactics. Common folklore about the 1965 war between India and Pakistan has it that the Pakistani soldiers were so virtuous (*naik*) that when Indian aircraft tried to destroy the Ravi Bridge (connecting Central and Northern Pakistan), divine help appeared in the form of *Sabz Imama Posh* (people wearing green *imamas*; *imama* is the traditional Arab male dress). These *Imama Poshs* would catch the falling bombs in their hands (like a cricket ball) and thus prevented the destruction of the bridge. However, a retired brigadier of the Pakistan Army who was deployed in Lahore during the 1965 war told me in an interview that "he wish(ed) it was the case, it would have been much easier for us" (field notes). The issue of *Sabz Imama Poshs* once again adorned the newspapers in Zia ul Haq's time in late 1980s.

5. The text lists selective Ulema who were not at the political forefront during the Pakistan Movement. Those listed include: Maulana Ashraf Ali Thanavi, Maulana Shabbir Ahmad Usmani, Maulana Abdul Hamid Badayuni, Maulana Zafar Ahmad Usmani, and Hafiz Kifayat Hussain (PTB 2002g: 152, PTB n.d.e: 47–48). The text omits to mention that Maulana Abul Kalam Azad was a staunch Congressite and that Maulana Maudoodi, the head of an influential religio-political party, vehemently opposed the idea of Pakistan.

6. Interestingly the discourse also articulates the masculine spaces. For example, on a number of occasions during my fieldwork in Pakistan I was asked why I wear my hair in a ponytail (like a woman). At other times, I was just ogled at or just laughed at. The most interesting case was when two young boys who had just broken their daylong Ramadan fast minutes ago, on seeing me walk by in Saddar, Rawalpindi, forgot all about eating and started yelling "golden golden, pony, pony." On at least two occasions, I was stopped by total strangers in the marketplace and was "advised" that I should "order" my wife (who was accompanying me) to cover her head and wear long sleeves. It also took a while for the students that I interviewed to get used to the fact that "uncle" (me) wore his hair in a ponytail. One nine-year-old, after three days of pondering, finally asked me if all men in Canada wore ponytails. They were shattered when I told them that the Prophet (PBH) also wore his hair long though not in a ponytail or a braid. It was obvious that for them a rather feminine image of the Prophet (PBH) amounted to blasphemy.

7. The essay on Ayisha portrays her as an obedient daughter and a good wife. It does not tell the student that she was an extremely spirited woman asked the Prophet (PBH) most questions and explanation of issues affecting women and is one of the major sources of Hadith. It also does not inform the students that Ayisha, after the death of the Prophet (PBH), challenged some of his companions (notably Hurayra) for passing on inauthentic Hadith. Nor does the essay tell the students that Ayisha challenged the fourth Pious Caliph Ali Ibne Abi Talib for state power and even lead in battles (for details see Mernissi 1987).

8. These figures are tabulated on the basis of full articles, essays, or poems on or about certain personalities. The presence of males in leadership positions otherwise is much more. Males in military leadership positions are not included in this section and are discussed separately. Male figures in leadership positions (religious and military) appear in the texts in the following order: Caliph Ali, Mohammad Iqbal (the national poet of Pakistan), and Sultan Bahoo (Sufi poet and mystic)—Urdu for class 2; Prophet Abraham (AS), caliphs Omar bin Khattab and Ali, and religious scholar Sheikh Saadi—class 3; Sufi poets and mystics Mian Muhammad Baksh and Waris Shah—class 4; Caliph Usman Ghani and Ottoman Caliph Haroon ur Rasheed—class 5; Companion of the Prophet Sa'ad bin Abi Waqas, Sufi Data Ganj Baksh, and nationalist leader Maulana Zafar Ali—class 6; Caliph Omer bin Khattab, Muslim general Khalid bin Waleed (conqueror of Spain), Farabi (Muslim scientist), and nationalist leaders Khawaja Nazim uddin, Sardar Abdul Rab Nishtar, and Maulvi Abdul Haq—class 7; and finally, Prophet's (PBH) grandson Hussain, Companion of the Prophet Bilal, religious scholar Bahuddin Zakria, Muslim scientist Ibnul Hasham, religious scholar Suleman Nadvi, and nationalist poets Altaf Hussain Hali and Muhammad Iqbal—class 8. In the new textbooks under the 2004 reform, a similar representational articulation can be noted. For example: Hazrat Ali Ibne Talib, Imam Abu Hanifa, Syed Abdul Qadir Jillani, and M. A. Jinnah—Urdu for class 2; Hazrat Omar bin Khitab, M. A. Jinnah, and Mohammad Iqbal—Urdu for class 3; Mohammad Iqbal, Data Ganj Bux, and Mian Mohammad Bux—Urdu for class 4; Jamaluddin Afghani, Sardar Abdul Rab Nishtar, Imam Jaffar, Hazrat Usman Ghani,

Naseerudding Zangi, Mahmud Ghaznavi, and Sachal Sarmast—Urdu for class 5; Maulvi Nazir Ahmad, Ali Hajveri, Jabir Bin Hayan, Liaqat Ali Khan, Mohammad Iqbal, and Salman Farsi—Urdu for class 6; Maulvi Abdul Haq, Farid Shakar Ganj, Akbar Illahabadi, and Mohammad Iqbal—Urdu for class 7; Salman Nadvi, M. A. Johar, M. A. Jinnah, Hazrat Bilal, Ibne Sina, Hasrat Mohan, and Baha uddin Zakria—Urdu for class 8. Other than these, religious and nationalist figures of various statures are sprinkled liberally all over the curricula and textbooks. Rampant in the presence of men in the texts are connotations of power and knowledge as exclusively male domains.

9. *Nishan e Haider* is the top military award of Pakistan. It is named after Ali (one of whose titles was *Haider e Karar*) the fourth of the Pious Caliphs known for his bravery.

10. Just as the stories of battles between India and Pakistan and those who fought with bravery and sacrificed their lives (and received *Nishan e Haider*) abound in the text, so do the narratives of *Ghazvat*. A list of full-essay narratives of *Ghazvat* includes Badr—Urdu for classes 5 and 6 (2002 and 2004), Yarmouk and Ohad—Urdu for class 7 (2002), and Karbala—Urdu for class 8 (2002). Other stories include military heroics of Khalid bin Waleed with respect to the Battle for Spain and other battles—Urdu for class 7 (2002 and 2004), Ma'az, Mouooz, and Hamza—Urdu for class 6 (2002).

11. Captain Mohammad Sarwar—Urdu for class 3 (Punjab Board), Lance Naik and Mohammad Mahfooz—Urdu for class 3 (2002), Rashid Minhas—Urdu for class 4 (2002 and 2004) and Urdu for class 6 (2004), Sawar Mohammad Hussain—Urdu for class 5 (2002) and Urdu for class 7 (2004), Major Tufail—Urdu for class 6, Sarwar Shaheed and Lance Naik Lal Hussain—Urdu for class 7 (2002 and 2004), and Lalak Jan—Urdu for class 3 (2004). Other than the ones listed above the list includes Major Raja Aziz Bhatti, Major Mohammad Akram, Major Shabbir Sharif, Hawaldar Lalak Jahan, and Captain Colonel Sher Khan. It is interesting to note the name of the last mentioned. The second word in the name, *Colonel*, does not designate his rank but is his middle name. It is a common tradition in some areas of Pakistan from where traditionally men are recruited for the armed forces to name sons after ranks or attributes such as *Bahadar* (brave), *Shaheed* (martyr), *Ghazi* (victorious), *Mujahid* (Islamic fighter), etc. This is one example of the normalcy of militarism.

12. Nazir Ahmad was a deputy collector for the British government in India and is known by both salutations.

# 8. Education and Gendered Citizenship in Pakistan

1. MMA stands for *Mutahida Majlis e Aml*, a coalition of religio-political parties of Pakistan. By virtue of its electoral gains in the 2002 election that it fought from an anti-US platform, this coalition has become a major parliamentary force in Pakistan.

# References

Abbas, M. (n.d.). *Textbook development in Pakistan and United Kingdom.* Lahore: Sang e Meel.

Abraham, I. (1995). Towards a reflexive South Asian security studies. In M. Weinbaum and C. Kumar (Eds.), *South Asia approaches millennium: Reexamining national security.* Boulder, CO: Westview Press.

Adeel, U., and M. Naqvi. (1997). *Pakistani women in development: A statistical mirror.* Academy Town Peshawar: Pakistan Academy for Rural Development.

Afzal, R. (1976). *Political parties in Pakistan: 1947–1958.* Islamabad: Quaid i Azam University Press.

Agger, B. (1992). *Cultural studies as critical theory.* London: Falmer Press.

Alavi, H. (1972). The state in post-colonial societies: Pakistan and Bangladesh. *New Left Review* 74.

———. (1983). State and class in Pakistan. In H. Gardezi and J. Rashid (Eds.), *Pakistan, the roots of dictatorship: The political economy of a praetorian state.* London: Zed Press.

———. (1986). Ethnicity, muslim society and the Pakistani ideology. In A. Weiss (Ed.), *Islamic reassertion in Pakistan: The application of Islamic laws in a modern state.* Syracuse: University of Pennsylvania Press.

———. (1990). Authoritarianism and legitimation of state power in Pakistan. In S. K. Mitra (Ed.), *The post-colonial state in Asia: Dialectics of politics and culture.* New York: Harvester/Wheatsheaf.

———. (1990a). The origins and significance of Pakistan-US military alliance. In S. Kumar (Ed.), *Yearbook on India's foreign policy.* New Delhi: Sage Publications.

———. (1991). Nationhood and communal violence in Pakistan. *Journal of Contemporary Asia* 21 (2): 152–78.

Alcoff, L. (1997). The politics of post-modern feminism revisited. *Cultural Critique* 36: 5–27.

Alexander, J., and C. Mohanty. (Eds.). (1997). *Feminist genealogies, colonial legacies, democratic futures.* New York: Routledge.

Ali, M. (1993). *In the shadow of history.* Lahore: Progressive.

———. (1993a). *Tareekh aur aurat* (History and Woman, Urdu). Lahore: Fiction House.

Allen, J. (1990). Does feminism need a theory of the state? In S. Watson (Ed.). *Playing the state: Australian feminist intervention*. London: Verso.

Alvarez, S. (1990). *Endangering democracy in Brazil: Women's movements in transition politics*. Princeton, NJ: Princeton University Press.

Amin, T. (1988). *Ethno-national movements of Pakistan: Domestic and international factors*. Islamabad: Institute of Policy Studies.

Anderson, B. (1983). *Imagined communities: Reflections on the origin and spread of nationalism*. London: Verso.

Anwar, M. (1982). *Images of male and female roles in school and college textbooks*. Islamabad: Women's Division, Government of Pakistan.

Appadurai, A. (1993). Patriotism and its future. *Public Culture* 5: 411–29.

Apple, M. (1990). *Ideology and curriculum*. New York: Routledge.

———. (1995). *Education and power*. New York: Routledge.

Apple, M., and L. Christian-Smith. (Eds.). (1991). *The politics of the textbook*. New York: Routledge.

Arnot, M., and G. Weiner. (Eds.). (1987). *Gender and the practice of schooling*. London: Hutchinson.

Ashcroft, B., G. Griffith, and H. Tiffin. (Eds.). (1995). *The post-colonial studies reader*. London: Routledge.

Assiter, A. (1996). *Enlightened women: Modernist feminism in a post-modern age*. London: Routledge.

Ayer, A. (1959). *Logical positivism*. London: Allen & Unwin.

———. (1975). *Central question of philosophy*. London: Weidenfeld & Nicholson.

———. (1985). *Philosophy in the twentieth century*. New York: Random House.

Azfar, K. (1991). Constitutional dilemmas. In S. J. Burki & C. Baxter (Eds.), *Pakistan under the military: Eleven years of Zia ul Haq*. Boulder, CO: Westview Press.

Azim, S. (1951). *Nai kitab* for class 6. Lahore: Urdu Markaz.

Aziz, K. (1993). *The murder of history: A critique of history textbooks used in Pakistan*. Lahore: Vanguard.

Bain, W. (1997). The loss of innocence: Lyotard, Foucault and the challenge of post-modern education. In M. Peters (Ed.), *Education and the post-modern condition*. Westport, CT: Bergin and Garvey.

Ball, S. (1990). Introducing monsieur Foucault. In S. Ball (Ed.), *Foucault and education: Discipline and knowledge*. London: Routledge.

Balochistan Textbook Board (2000). *Meri kitab for class 3*. Quetta: New Nisa Printers.

Barrett, M. (1980). *Women's oppression today: Problems in Marxist feminist analysis*. London: Verso.

———. (1989). *Women's oppression today: The Marxist/feminist encounter*. London: Verso.

Bartky, S. (1988). Foucault, femininity and the post-modernization of patriarchal power. In I. Diamond and L. Quinby (Eds.), *Feminism and Foucault: Reflections on resistance*. Boston: Northeastern University Press.

Bell, D., and R. Klein. (Eds.). (1996). *Radically speaking: Feminism reclaimed*. London: Zed Books.

Benhabib, S. (1992). *Situating the self*. New York: Routledge.

————. (1994). From identity politics to social feminism: A plea for the nineties. *PES Yearbook,* 1994. http://www.ed.uiuc.edu.

Bertaux, D. (Ed.). (1981). *Biography and Society: The life history approach in social sciences.* Beverly Hills, CA: Sage Publications.

Beyer, L., and M. Apple. (Eds.). (1988). *The curriculum: Problems, politics, and possibilities.* Albany: State University of New York Press.

Bhaba, H. (1990). The third space: Interview with Homi Bhaba. In J. Rutherford (Ed.), *Identity: community, culture, difference.* London: Lawrence & Wishart.

————. (1994). *Location of culture.* London: Routledge.

Bi'llah, M. (1953). *Shahreeat* (Civics). Part I for class 6. Sixth Edition. Lahore: Publishers United.

Bjerg, J., and H. Silverbrandt. (1980). Roskilde University Center—A Danish experiment in higher education. *Curriculum Studies* 12 (3): 245–61.

Black, L. (2001). The predicament of identity. *Ethnohistory* 48 (1–2): 337–50.

Blades, D. (1997). *Procedures of power and curriculum change: Foucault and the quest for possibilities in science education.* New York: Peter Lang.

Blum, L. (1999). Race, community and moral education: Kohlberg and Spielberg as civic educators. *The Journal of Moral Education* 28 (2): 201–9.

Bose, S., and A. Jalal. (1998). *Modern South Asia: History, culture, political economy.* Lahore: Sang-e-Meel Publications.

Boston, P., S. Jordan, E. McNamara, and K. Kozolanka. (1997). Using participatory action research to understand the meanings aboriginal Canadians attribute to the rising incidence of diabetes. *Canada* 18: 5–12.

Bourdieu, P. (1990). *In other words: Essays towards a reflexive sociology.* Translated by Matthew Adamson. Oxford: Polity.

————. (1990a). *An Introduction to the work of Pierre Bourdieu: The practice of theory.* R. Harker, C. Mahar, and C. Wilkes (Eds.). New York: St. Martin's Press.

Brah, A., and R. Minhas. (1985). Structural racism or cultural difference: Schooling for Asian girls. In G. Weiner (Ed.), *Just a bunch of girls: Feminist approaches to schooling.* Milton Keynes, UK: Open University Press.

Brass, P. (1979). Elite groups, symbol manipulation and ethnic identity among the muslims of South Asia. In D. Taylor and M. Yepp (Eds.), *Political identity in South Asia.* London: Curzon Press.

Brown, A. (2000). *On Foucault.* Belmont, CA: Wadsworth Thomson Learning.

Brown, G., and G. Yule. (1983). *Discourse analysis.* Cambridge: Cambridge University Press.

Brownmiller, S. (1977). *Against our will.* Harmondsworth, UK: Penguin.

Bryson, V. (1995). Adjusting the lenses: Feminist analysis and Marxism at the end of the twentieth century. *Contemporary Politics* 1 (1).

————. (1999). *Feminist debates: Issues of theory and political practice.* New York: New York University Press.

Bubeck, D. (1995). *Care, gender and justice.* Oxford: Clarendon Press.

Burchell, G. et al. (1991). *The Foucault effect: Studies in governmentality: With two lectures and an interview with Michel Foucault.* Chicago: University of Chicago Press.

Butler, J. (1990). *Gender trouble: Feminism and the subversion of identity.* London: Routledge.

Butler, M. (1995). Early liberal roots of feminism: John Locke and the attack on patriarchy. In N. Tuana and R. Tong (Eds.), *Feminism and philosophy.* London: Westview Press.

Caldwell, C. (2009) *Reflections on the Revolution in Europe. Immigration, Islam and the West.* New York: Doubleday.

Calhoun, C. (1991). The problem of identity in collective action. In J. Huber (Ed.), *Macro-micro linkages in sociology.* Beverly Hills, CA: Sage.

———. (1994). Social theory and the politics of identity. In C. Calhoun (Ed.), *Social theory and the politics of identity.* Oxford: Blackwell.

———. (1995). *Critical social theory.* Oxford: Blackwell.

Calinescu, M. (1985). Introductory remarks. In M. Calinescu and D. Fokkema (Eds.), *Exploring post-modernism.* Amsterdam: John Benjamin Publishing Company.

Carby, H. (1987). Black feminism and the boundaries of sisterhood. In M. Arnot and G. Weiner (Eds.), *Gender and the practice of schooling.* London: Hutchinson.

Cascardi, A. (1992). *The subject of modernity.* New York: Cambridge University Press.

Charlton, S., J. Everett, and K. Staudt. (Eds.). (1989). *Women, the state and development.* Albany: State University of New York Press.

Chatterjee, P. (1986). *Nationalist thought and the colonial world: A derivative discourse?* London: Zed Books.

———. (1993). *The nation and its fragments: Colonial and postcolonial histories.* Princeton, NJ: Princeton University Press.

———. (1994). Secularism and toleration. *Economic and Political Weekly* 29 (28): 1786–97.

Cherryholmes, C. (1987). A social project for curriculum: Post-structural perspectives. *Journal of Curriculum Studies* 20 (1): 1–21.

———. (1988). *Power and criticism: Poststructural investigations in education.* New York: Teachers College Press.

Choudhury, G. (1963). *Democracy in Pakistan.* Dacca: Green Book House.

———. (1969). *Constitutional development in Pakistan.* Harlow, UK: Longmans.

Choudhury, G. W. (1988). *Pakistan: Transition from military to civilian rule.* Buckhurst Hill, UK: Scorpion Publishing.

Chouliaraki, L., and N. Fairclough. (1999). *Discourse in late modernity. Rethinking critical discourse analysis.* Edinburgh: Edinburgh University Press.

Collins, P. (1990). *Black feminist thought: Knowledge, consciousness, and the politics of empowerment.* New York: Routledge.

———. (1998). *Fighting words: Black women and the search of justice.* Minneapolis: University of Minnesota Press.

———. (2000). What's going on? Black feminist thought and the politics of postmodern. In E. St. Pierre and W. Pillow (Eds.), *Working the ruins: Feminist poststructural theory and methods in education.* New York: Routledge.

Combahee River Collective (1982). A black feminist statement. In G. T. Hull et al. (Eds.), *All the women are white, all the blacks are men, but some of us are brave: Black women's studies.* New York: Feminist Press.

Connell, R. (1988). *Gender and power.* Sydney: Allen and Unwin.

Connell, R.W. (1987) Gender and Power. Sydney: Allen & Unwin.

———. (1990). The state, gender and sexual politics. *Theory and Society* 19 (5): 501–44.

Connolly, W. (1991). *Identity/difference: Democratic negotiations of political paradox.* Ithaca, NY: Cornell University Press.

Counihan, T. (1976). Epistemology and science: Feyerabend and Lecourte. *Economy and Society* 5 (1).

Cousins, M., and A. Hussein. (1984). *Michel Foucault.* London: Macmillan.

Covaleski, J. (1993). Power goes to school: Teachers, students, and discipline. *Philosophy of Education Society Yearbook.* http://www.Ed.uiuc.edu/EPS/PES-Yearbook/93_docs/COVALESK.htm.

Curle, A. (1973). *Educational problems of developing societies with case studies of Ghana, Pakistan and Nigeria.* New York: Praeger.

Curtis, B. (1988). Policing pedagogical space: Voluntary school reform and moral regulation. *Canadian Journal of Sociology* 13 (3): 283–304.

Da Silva, T. (1988). Distribution of school knowledge and social reproduction in a Brazilian urban setting. *British Journal of Sociology of Education* 9 (1): 55–79.

Dahlerup, D. (1987). Confusing concepts—confusing reality: A theoretical discussion of the patriarchal state. In A. Showstock-Sassoon (Ed.), *Women and the state: Shifting the boundaries of the public and the private.* London: Hutchinson.

Daly, M. (1978). *Gyn/ecology: The metaethics of radical feminism.* Boston: Beacon Press.

Davies, B. (1992). Poststructuralist theory and critical practice. Geelong, AU: Deakin University Press.

———. (1992a). Women's subjectivity and feminist stories. In C. Ellis and M. Flaherty (Eds.), *Investigating subjectivity: Research on lived experience.* London: Sage.

Davies, B., and C. Banks. (1992). The gender trap: A feminist post-structuralist analysis of primary-schools children's talk about gender. *Journal of Curriculum Studies* 24 (1): 1–25.

Davis, A. (1982). *Women, race and class.* London: Women's Press.

———. (1990). *Women, culture and politics.* London: Women's Press.

Davis, M. (1995). Towards a theory of Marxism and oppression. *Contemporary Politics* 1 (2).

*Dawn* (Karachi). November 27, 1991.

*Dawn Daily* (Internet Edition). March 12, 2004. http://www.dawn.com.

De Beauvoir, S. (1965). *Second sex.* New York: Penguin.

Deleuze, G. (1986). *Foucault.* Minneapolis: University of Minnesota Press.

Denzin, N, and Y. Lincoln. (Eds.). (1998). *Strategies of qualitative inquiry.* Thousand Oaks, CA: Sage Publishers.

Denzin, N. (1997). *Interpretive ethnography: Ethnographic practices for the 21st century.* Thousand Oaks, CA: Sage Publishers.

Department of Education. (1956). *Urdu ki Satween Kitab*. Lahore.

Dewey, C. (1991). The rural roots of Pakistani militarism. In D. A. Low (Ed.), *The political inheritance of Pakistan*. New York: St. Martin's Press.

Dietz, M. (1987). Context is all: Feminism and the theories of citizenship. *Daedalus* 116: 1–24.

Dillard, C. (2000). The substance of things hoped for, the evidence of things not seen: Examining an endarkened feminist epistemology in educational research. *Qualitative Studies in Education* 13 (6): 661–81.

Dirks, N. (1992). Introduction. In N. Dirks (Ed.), *Colonialism and culture*. Ann Arbor: University of Michigan Press.

Dirlik, A. (1994). The post-colonial aura: Third world criticism in the age of global capitalism. *Critical Inquiry* 20 (2): 328–56.

Downes, W. (1988). *Language and society*. London: Fontana.

Dreyfus, H., and P. Rabinow. (Eds.). (1983). *Michel Foucault: Beyond structuralism and hermeneutics*. Chicago: University of Chicago Press.

Du Preez, P. (1980). *The problem of identity*. New York: St. Martin Press.

Dworkin, A. (1981). *Pornography: Men possessing women*. London: Women's Press.

Dyson, M. (1993). *Reflecting black: African American cultural criticism*. Minneapolis: University of Minnesota Press.

Economic Survey of Pakistan. (2003). http://www.finance.gov.pk/survey/home/htm.

Edwards, D., and D. Potter. (1992). *Discursive psychology*. Newbury Park, CA: Sage.

Eisenstein, Z. (1981). *The radical future of liberal feminism*. London: Longman.

———. (1995). The sexual politics of the new right: Understanding the 'crisis of liberalism' for the 1980s. In. N. Tuana and R. Tong (Eds.), *Feminism and philosophy*. London: Westview Press.

Escobar, A., and S. Alvarez. (Eds.). (1992). *The making of social movements in Latin America: Identity, strategy and democracy*. Boulder, CO: Westview.

Esposito, J. (1987). *Islam in Asia: Religion, politics, and society*. New York: Oxford.

———. (1994). Islam, ideology and politics in Pakistan. In M. Weiner and A. Banuazizi (Eds.), *The politics of social transformation in Afghanistan, Iran and Pakistan*. Syracuse, NY: Syracuse University Press.

———. (1996). *Islam and democracy*. New York: Oxford University Press.

———. (1998). *Islam and politics*. Syracuse, NY: Syracuse University Press.

Etzioni, A. (1993). *The spirit of community*. New York: Crown.

Evans, J. (1995). *Feminist theory today: An introduction to second wave feminism*. London: Sage.

Evans, P., D. Rueschemeyer, and T. Skocpol. (Eds.). (1985). *Bringing the state back*. Cambridge: Cambridge University Press, 1985.

Ewing, K. P. (Ed.). (1988). *Shariat and ambiguity in South Asian Islam*. Berkeley: University of California Press.

Fairclough, N. (1992). *Discourse and social change*. Cambridge, MA: Polity Press.

———. (1995). *Critical discourse analysis: The critical study of language*. London: Longman.

———. (1995a). *Media discourse*. London: E. Arnold.

Ferguson, J. (1990). *The anti-politics machine: Development, depoliticization, and bureaucratic power in Lesotho*. Cambridge: Cambridge University Press.

Ferre, F. (1996). *Being and value: Toward a constructive post-modern metaphysics*. Albany: State University of New York Press.

Firestone, S. (1979). *The dialectic of sex*. London: Women's Press.

Flax, J. (1990). Post-modernism and gender relations in feminist theory. In L. Nicholson (Ed.), *Feminism/post-modernism*. New York: Routledge.

Foucault, M. (1973). *The order of things: An archeology of human sciences*. London: Tavistock.

———. (1977). *The archaeology of knowledge*. London: Tavistock.

———. (1977a). *Language, counter-memory, practice: Selected essays and interviews*. Edited by D. F. Bouchard. Translated by D. Bouchard and S. Simon.. Ithaca: Cornell University Press.

———. (1978). *I, Pierre Riviere, having slaughtered my mother, my sister, and my brother: A case of parricide in the 19th century*. Edited by M. Foucault. Translated by F. Jellinek. Lincoln: University of Nebraska Press.

———. (1978a). Politics and the study of discourse. *Ideology and Consciousness* 3: 7–26.

———. (1979). *Discipline and punish: The birth of the prison*. New York: Vintage Books.

———. (1979a). On governmentality. *Ideology and Consciousness 6*.

———. (1980). *Power/knowledge: Selected interviews and other writings 1972–1977*. Edited by C. Gordon. New York: Pantheon.

———. (1980a). Introduction. *Herculine Barbin: Being the recently discovered memoirs of a nineteenth-century French hermaphrodite*. Translated by R. McDougall, Trans. New York: Pantheon Books.

———. (1982). *This is not a pipe*. Translated by James Harkness. Los Angeles: University of California Press.

———. (1982a). *Death and the labyrinth: The world of Raymond Roussel*. Translated by C. Ruas. Berkeley: University of California Press.

———. (1990). *Politics, philosophy culture: Interviews and other writings 1977–1984*. Edited by L. D. Kritzman. Translated by A. Sheridan and others. New York: Routledge.

———. (1990a). *The history of sexuality*. Volume 2: *The use of pleasure*. Translated by R. Hurley. New York: Vintage Books.

———. (1990b). *The history of sexuality*. Volume 1: *An introduction*. Translated by R. Hurley. New York: Vintage Books.

———. (1994). *The Birth of the clinic: An archeology of medical perception*. Translated by A. M. Sheridan. New York: Vintage.

Foweraker, J. (1995). *Theorizing social movements*. Boulder, CO: Pluto Press. 1995.

Franzway, S. (1986). With problems of their own: Femocrats and the welfare state. *Australian Feminist Studies 3*.

Fraser, N., and L. Nicholson. (1990). Social criticism without philosophy: An encounter between feminism and postmodernism. In L. Nicholson (Ed.), *Feminism/postmodernism*. London: Routledge.

Freedman, K., and T. Popkewitz. (1988). Art education and social interests in the development of American schooling: Ideological origins of curriculum theory. *Journal of Curriculum Studies* 20 (5): 387–405.

Friedan, B. (1986). *The feminine mystique.* Harmondsworth, UK: Penguin Books.

Frondizi, R. (1971). *What is value? An introduction to axiology.* Translated by S. Lipp. La Salle, IL: Open Court.

Fuss, D. (1989). *Essentially speaking feminism, nature and difference.* New York: Routledge.

Gardezi, F. (1994). Islam, feminism and the women's movement in Pakistan: 1981–1991. In K. Bhasin, R. Menon and N. S. Khan (Eds.), *Against all odds: Essays on women, religion and development from India and Pakistan.* New Delhi: Kali for Women.

Garfinkel, H. (1963). A conception of and experiments with "trust as a condition of stable concerted efforts". In O. Harvey (Ed.), *Motivation and social interaction.* New York: Ronald Press.

———. (1984). *Studies in ethnomethodology.* Oxford: Blackwell.

*Gazette* (Montreal). March 21, 2004, p. 2.

Geertz, C. (1973). *The interpretation of culture.* New York: Basic Books.

———. (1983). *Local knowledge: Further essays in interpretive anthropology.* New York: Basic Books.

Gergen, K. (1995). Social construction and the transformation of identity politics. *New School for Social Research Symposium.*

Ghosh, R. (1981) Minority within a minority: On being South Asian and female in Canada. In G. Kurian and R. Ghosh (Eds.), *Women in the family and economy: An international comparative survey.* Westport, CT: Greenwood Press.

Ghosh, R. (2002). *Redefining multicultural education.* 2nd edition. Toronto: Nelson Thomson Learning.

Ghosh, R., and M. A. Naseem. (In Press). *Colonial and post-colonial construction of identity: A comparison of teenage identity in Macau, Hong Kong and Goa.* Islamabad: PICRA Research Series Monograph.

Ghosh, R., A. Abdi, and M. A. Naseem. (Forthcoming). Postcolonial theories of Identity: Review of some current debates and theorization.

Gillin, T. (1993). The rise of identity politics. *Dissent* 40: 172–77.

Gaborieau, M. (2002). Religion in the Pakistani polity. In S. Mumtaz, J. Racine, and I. A. Ali (Eds.), *Pakistan: The contours of state and society.* Karachi: Oxford University Press.

Goffman, E. (1963). *Behavior in public places.* New York: Free Press.

———. (1969). *The presentation of self in everyday life.* Harmondsworth, UK: Penguin.

Goody, J. (1976). *Production and reproduction: A comparative study of the domestic domain.* Cambridge: Cambridge University Press.

———. (1983). *The development of the family and marriage in Europe.* Cambridge: Cambridge University Press.

Government of Pakistan. (1947). Ministry of the Interior (Education Division). *Proceedings of the Pakistan educational conference.*

———. (1951). Education Division. *Proceedings of the educational conference.*

Government of Pakistan. (1957). National Planning Board. *First five-year plan: 1955–60.*

———. (1959). Ministry of Education. *Report of the commission on national education.*

———. (1970). Ministry of Education and Scientific Research. *The new education policy.*

———. (1972). Ministry of Education. *The education policy.*

———. (1979). Ministry of Education. *National education policy and implementation programme.*

———. (1981). Ministry of Education. *The gazette of Pakistan, Part 111, February 17.*

———. (1983). *Yasarnal Quran* (Quranic Primer).

———. (1985). *Report of the commission on the status of women.* Islamabad: Printing Corporation of Pakistan.

———. (1985a). Literacy and Mass Education Commission. *PC-1, National literacy programme: 1984–85.*

———. (1986). Literacy and Mass Education Commission. *PC-1, Drop-in schools.*

———. (1986a). Literacy and Mass Education Commission. *PC-1, Nationwide literacy programme.*

———. (1986b). Literacy and Mass Education Commission. *PC-1, Nai Roshni schools.*

———. (1989). Academy of Educational Planning and Management. *Evaluation of Iqra pilot project.*

———. (1992). Ministry of Education. *National education policy.*

Government of Pakistan, Federal Ministry of Education (Curriculum Wing). (2002). *English* (compulsory) for classes 1–5. Islamabad: National Book Foundation.

———. (2002a). *English* (compulsory) for classes 6–8. Islamabad: National Book Foundation.

———. (2002b). *Early childhood education.* Islamabad: National Book Foundation.

———. (2002c). *Marboot nisab* (integrated curriculum) for classes 1–3 (Urdu). Islamabad: National Book Foundation.

———. (2002d). National Curriculum. *Social studies* for classes 1–5. Islamabad: National Book Foundation.

———. (2002e). National Curriculum. *Social studies* for classes 6–8. Islamabad: National Book Foundation.

———. (2002f). *National curriculum: Education* for classes 9–10. Islamabad: National Book Foundation.

———. (2002g). *National curriculum: Social studies* for classes 11–12. Islamabad: National Book Foundation.

———. (2002h). *Nisab Islamyat* (Islamyat curriculum) for classes 1–5 (Urdu). Islamabad: National Book Foundation.

———. (2002i). *Nisab Islamyat* (Islamyat curriculum) for classes 6–7 (Urdu). Islamabad: National Book Foundation.

———. (2002j). *Nisab Islamyat* (Islamyat curriculum) for classes 9–10 (Urdu). Islamabad: National Book Foundation.

Government of Pakistan. (2002k). *Nisab Urdu: Zaban Awal* (Urdu curriculum: First language) for classes 4–5 (Urdu). Islamabad: National Book Foundation.

———. (2002l). *Nisab Urdu: Zaban Awal* (Urdu curriculum: First language) for classes 6–7 (Urdu). Islamabad: National Book Foundation.

———. (2002m). *Nisab Urdu: Zaban Awal* (Urdu curriculum: First language) for classes 11–12 (Urdu). Islamabad: National Book Foundation.

———. (2002n). *Nisab Urdu: Lazmi* (Urdu curriculum: Compulsory) for classes 11–12 (Urdu). Islamabad: National Book Foundation.

———. (2002o). *Urdu ka naya qaida* (primer) for class 1. Islamabad: National Book Foundation.

Government of Pakistan, Federal Ministry of Education. (2001). *Pehli darsi kitab* for class 1. Islamabad: National Book Foundation.

———. (2001a). *Doosri darsi kitab* for class 2. Islamabad: National Book Foundation.

———. (2001b). *Teesri darsi kitab* for class 3. Islamabad: National Book Foundation.

———. (2001c). *Teesri darsi kitab* for class 3. Experimental edition, vol. 1. Islamabad: National Book Foundation.

———. (2001d). *Teesri darsi kitab* for class 3. Experimental edition, vol. 2. Islamabad: National Book Foundation.

Government of Pakistan, Ministry of Education website. http://www.moe,gov.pk.

Government of Pakistan, Planning Commission. (1978). *Fifth five-year plan: 1978–83.*

———. (1983). *Sixth five-year plan: 1983–88.*

———. (1988). *Seventh five-year plan: 1988–93.*

———. (1993). *Eighth five-year plan: 1993–98.*

———. (1998). *Ninth five-year plan: 1998–03.*

Government of Pakistan, Statistics Division. http://www.statpak.gov.pk/depts/az/az.html.

Greenhalgh, S. (1992). Negotiating birth control in village China. *Population Council Working Papers* 38.

———. (1992a). The changing value of children in the transition from socialism: The view from three Chinese villages. *Population Council Working Papers* 43.

———. (1994). The social dynamics of child mortality in village Shaanxi. *Population Council Working Paper* 66.

Greenhalgh, S., and L. Jalai. (1995). Endangering reproductive policy and practice in peasant China: For a feminist demography of reproduction. *Signs* 20.

Greer, G. (1971). *The female eunuch.* London: Paladin.

Gubrium, J., and J. Holstein. (2002). *Handbook of interview research: Context & method.* Thousand Oaks, CA: Sage Publications.

Guha, R., and G. Spivak. (Eds.). (1988). *Selected subaltern studies.* New York: Oxford University Press.

Hafeez, S. (1981). *The metropolitan women in Pakistan.* Karachi: Royal Book Co.

Hall, S. (1985). Signification, representation, ideology: Althusser and the post-structuralist debates. *Critical Studies in Mass Communication* 2 (2): 91–114.

Hammersley, M. (1992). *What's wrong with ethnography?* London: Routledge.

Haq, F. (1996). Women, Islam and the state in Pakistan. *Muslim World* 86 (2): 158–75.

Haraway, D. (1991). Situated knowledges: The science question in feminism and the privilege of partial presence. In D. Haraway (Ed.), *Simians, cyborgs and women: The reinvention of nature.* New York: Routledge

Harding, S. (1990). Feminism, science, and the anti-enlightenment critiques. In L. Nicholson (Ed.), *Feminism/post-modernism.* New York: Routledge.

Harstock, N. (1985). *Money sex and power: Towards a feminist historical materialism.* Boston: Northeastern University Press.

———. (1987). Rethinking modernism, majority versus minority theories. *Cultural Critique* 7: 187–206.

———. (1990). Foucault on power: A theory for women? In L. Nicholson (Ed.), *Feminism/post-modernism.* New York: Routledge.

Hartman, H. (1983). *The unhappy marriage of Marxism and feminism: Towards a more progressive union.* In L. Sargent (Ed.), *The unhappy marriage of Marxism and feminism.* London: Pluto Press.

Hasan, I. (1993). Gender analysis of middle school textbooks. *Journal of Behavioral Sciences* (June).

Hasnain, K., and A. H. Nayyar. (1997). Conflict and violence in the educational process. In Z. Mian and I. Ahmad (Eds.), *Making enemies, creating conflict: Pakistan's crises of state and society.* Lahore: Mashal.

Hekman, S. J. (1990). *Gender and knowledge: Elements of post-modern feminism.* Boston: Northeastern University Press.

Henriques, J., et al. *Changing the subject.* London: Methuen.

Hickling-Hudson, A. (1999) "Experiments in Political Literacy: Caribbean Women and Feminist Education." *Journal of Education and Development in the Caribbean* 3 (1): 19–44.

Hodder, I. (2000). The interpretation of documents and material culture. In N. Denzin and Y. Lincoln (Eds.), *Handbook of qualitative research.* 2nd edition. Thousand Oaks, CA: Sage.

Hoodbhoy, P. (Ed.). (1998). *Education and the state: Fifty years of Pakistan.* Karachi: Oxford University Press.

———. (1998a). Preface: Out of Pakistan's education morass: Possible? How? In P. Hoodbhoy (Ed.), *Education and the state: Fifty years of Pakistan.* Karachi: Oxford University Press.

Hoodbhoy, P. and A. Nayyar. (1985). Rewriting the history of Pakistan. In A. Khan (Ed.), *Islam, Pakistan and the state: The Pakistan experience.* London: Zed Books.

hooks, b. (1981). *Ain't I a woman: Black women and feminism.* Boston: South End Press.

———. (1984). *Feminist Theory: From margin to center.* Boston: South End Press.

———. (1988) *Talking back: Thinking feminist, thinking black.* Boston: South End Press.

Hoskins, K. (1990). Foucault under examination: The crypto-educationalist unmasked. In S. Ball (Ed.), *Foucault and education: Discipline and knowledge.* London: Routledge.

Hoy, D. (1986). *Foucault: A critical reader.* Oxford: Basil Blackwell.

Human Rights Commission of Pakistan. (2003). Annual Report. http://www. hrcp-web.org/.

Humm, M. (1987). *An annotated bibliography of feminist criticism.* Boston: G. K. Hall.

Hussain, A. (1963). *The educated Pakistani girl: A sociological study.* Karachi: Ima Printers.

Hussain, N. (1994). Women as objects and women as subjects within fundamentalist discourse. In N. S. Khan, R. Saigol and A. S. Bano (Eds.), *Locating the self: Perspectives on women and multiple identities.* Lahore: ASR.

Hussain, N., K. Mumtaz, and R. Saigol. (Eds.). (1997). *Engendering the nation-state.* Lahore: Simorgh Women's Resource and Publication Centre.

Hussain, N., and N. Shah. (1991). Women, media and the production of meaning. In F. Zafar (Ed.), *Finding our way: Readings on women in Pakistan.* Lahore: ASR.

Huyssen, A. (1990). Mapping the post-modern. In L. Nicholson (Ed.), *Feminism/post-modernism.* New York: Routledge.

Jaggar, A. (1983). *Feminist politics and human nature.* Brighton, UK: Harvester.

Jahan, R. (1972). *Pakistan: Failure in national integration.* New York: Columbia University Press.

Jahangir, A., and H. Jillani. (2003). *Hudood ordinances: A divine sanction?* Lahore: Rohtas Books.

Jalal, A. (1985). *The sole spokesman: Jinnah, the Muslim League, and the demand for Pakistan.* Cambridge: Cambridge University Press.

———. (1990). *The state of martial rule: The origins of Pakistan's political economy of defence.* Cambridge: Cambridge University Press.

———. (1990a). The convenience of subservience: Women and the ztate in Pakistan. In D. Kandiyoti (Ed.), *Women, Islam and the state.* London: Macmillan.

———. (1995). *Democracy and authoritarianism in South Asia: A comparative and historical perspective.* Cambridge: Cambridge University Press.

Jalil, N. (1998). Pakistan's education—The first decade. In P. Hoodbhoy (Ed.), *Education and the state: Fifty years of Pakistan.* Karachi: Oxford University Press.

Jamal, A. (1995). Identity, community and post-colonial experience of migrancy. *Resources for feminist research* 23 (4): 35–41.

———. (2002). Entangled modernities: Feminism, Islamization and citizenship in a transnational locale; Women and the Pakistan nation-state. Unpublished Doctoral Thesis. Department of Sociology and Equity Studies. Toronto: OISE, University of Toronto.

Jameson, F. (1981). *The political unconscious: Narrative as a socially symbolic act.* London: Methuen.

———. (1995). On cultural studies. In J. Rajchman (Ed.), *The identity question.* New York: Routledge.

*Jang Daily* (Lahore), March 18, 1982.

Jeffery, P. (1998). Agency, activism and agendas. In P. Jeffery and A. Basu, (Eds.), *Appropriating gender: Women's activism and politicized religion in South Asia.* New York: Routledge.

Jones, K. (1993). *Compassionate authority: Democracy and the representation of women.* London: Routledge.

Jordan, S., and D. Yeoman. (1995). Critical ethnography: Problems in contemporary theory and practice. *British Journal of Education* 16 (3): 389–408.

Jørgensen, M., and L. Phillips. (2002). *Discourse analysis as theory and method.* London: SAGE Publications.

Kardar, S. (1998). The economics of education. In P. Hoodbhoy (Ed.), *Education and the state: Fifty years of Pakistan.* Karachi: Oxford University Press.

Kazi, A. (1987). *Ethnicity and education in nation-building: The case of Pakistan.* Lanham, MD: University Press of America.

Kazi, S., and Z. Sathar. (1991). Women's roles: Health, education and reproductive behavior. In F. Zafar (Ed.), *Finding our way: Readings on women in Pakistan.* Lahore: ASR.

Kenway, J. (1995). Feminist theories of the state: To be or not to be? In M. Blair and J. Holland with S. Sheldon (Eds.), *Identity and diversity: Gender and the experience of education.* Clevedon, UK: Adelesh.

Khan, N. (Ed.). (1992). *Voices within: Dialogues with women on Islam.* Lahore: ASR.

Khan, N., R. Saigol, and A. Zia. (Eds.) (1994). *Locating the self: Perspectives on women and multiple identities.* Lahore: ASR.

Khan, N., A. Zia, and R. Saigol. (1994). *Unveiling the issues: Pakistani women's perspectives on social, political and ideological issues.* Transcribed and translated by Naureen Amjad and Rubina Saigol. Lahore: ASR.

Khattak, S. (1994). Militarization, masculinity and identity in Pakistan: Effects on women. In N. Khan, A. Zia, and R. Saigol. (1994), *Unveiling the issues: Pakistani women's perspectives on social, political and ideological issues.* Transcribed and translated by Naureen Amjad and Rubina Saigol. Lahore: ASR.

———. (1994a). A reinterpretation of the state and statist discourse in Pakistan: 1977–88. In N. S. Khan, R. Saigol and A. S. Bano (Eds.), *Locating the self: Perspectives on women and multiple identities.* Lahore: ASR.

Khory, K. (1997). The ideology of the nation-state and nationalism. In R. Rais (Ed.), *State, society and democratic change in Pakistan.* Karachi: Oxford University Press.

Kipinis, L. (1988). Feminism: The political conscience of postmodernism? In A. Ross (Ed.), *Universal abandon: The politics of postmodernism.* Minneapolis: University of Minnesota Press.

Kirby, S., and K. McKenna. (1989). *Experience, research, social change: Methods from the margins.* Toronto: Garamond Press.

Kurin, R. (1985). Islamization in Pakistan: A view from the countryside. *Asian Survey* 25 (8): 854–61.

Laclau, E., and C. Mouffe. (2001). *Hegemony and socialist strategy: Towards a radical democratic politics.* London: Verso.

Lakoff, G. (1987). *Women, fire, and dangerous things: What categories reveal about the mind*. Chicago: University of Chicago Press.

Lather, P. (1991). *Getting smart*. New York: Routledge.

———. (1993). Fertile obsession: Validity after poststructuralism. *Sociological Quarterly* 34: 673–94.

———. (2000). Drawing the line at angles: Working the ruins of feminist ethnography. In E. St. Pierre and W. Pillow (Eds.), *Working the ruins: Feminist poststructural theory and methods in education*. New York: Routledge.

Lefebvre, H. (1971). *Everyday life in the modern world*. London: Allen Lane.

Lievesely, G. (1996). Stages of growth? Women dealing with the state and each other in Peru. In S. Rai and G. Lievesely (Eds.), *Women and the state: International perspectives*. London: Taylor and Francis.

Liu, T. (1994). Teaching the differences among women from a historical perspective: Rethinking race and gender as social categories. In V. Ruiz and E. Dubois (Eds.), *Unequal sisters: A multicultural reader in US women's history*. London: Routledge.

Livingstone, E. (1987). *Making sense of ethnomethodology*. London: Routledge & Kegan Paul.

Loomba, A. (1998). *Colonialism/Postcolonialism*. London: Routledge.

Lorde, A. (1984) *Sister Outsider. Essays and Speeches*. New York: Crossing Press.

Luke, C., S. Castell, and A. Luke. (1983). Beyond criticism: The authority of the school text. *Curriculum Inquiry* 13 (2): 111–27.

Lydon, M. (1988). Foucault and feminism: A romance of many dimensions. In I. Diamond and L. Quinby (Eds.), *Feminism and Foucault: Reflections on resistance*. Boston: Northeastern University Press.

Lyon, D. (1999). *Postmodernity*. Minneapolis: University of Minnesota Press.

Lyotard, J. (1984). *The post-modern condition: A report on knowledge*. Translated by G. Bennington and B. Massumi. Minneapolis: University of Minnesota Press.

Macaulay, T. Lord. (1995). Minute on Indian education. In B. Ashcroft, G. Griffith, and H. Tiffin (Eds.). *The post-colonial studies reader*. London: Routledge.

Macherey, P. (1965). *A theory of literary production*. London: Routledge.

MacKinnon, C. (1983). Feminism, Marxism, method and the state: Towards feminist jurisprudence. *Signs* 8 (4).

———. (1989). Towards *a feminist theory of the state*. Cambridge, MA: Harvard University Press.

Maguire, M. H. (1999) A bilingual child's discourse choices and voices: Lessons in listening, noticing and understanding. In E. Franklin (Ed.), *Reading and writing in more than one language: Lessons from teachers*. teachers of English and speakers of other languages. Series Editor: David Nunan.

Malik, I. (1997). *State and civil society in Pakistan: Politics of authority, ideology, and ethnicity*. New York: St. Martin's Press.

Mama, A. (1989). Violence against black women: Gender, race and state responses. *Feminist Review* 32.

Manning, P. (1992). *Erwin Goffman and modern sociology.* Cambridge: Polity.

Marilley, S. (1996). *Woman suffrage and the origins of liberal feminism in the United States 1820–1920.* Cambridge, MA: Harvard University Press.

Marker, M. (1991). On patriarchy and Islam. In F. Zafar (Ed.), *Finding our way: Readings on women in Pakistan.* Lahore: ASR.

Markovits, C. (2002). Cross-currents in the historiography of partition. In S. Mumtaz, J. Racine, and I. Ali, (Eds.), *Pakistan: The contours of state and society.* Karachi: Oxford University Press.

Marshall, H., and M. Arnot (2008) "The Gender Agenda: The Limits and Possibilities of Global and National Citizenship Education." In Joseph Zajda, Lynn Davies and Suzanne Majhanovich, (Eds.), *Comparative and Global Pedagogies. Equity, Access and Democracy in Education.* Netherlands: Springer, pp. 103–24.

Marshall, J. (1990). Foucault and educational research. In S. Ball, (Ed.), *Foucault and education: Discipline and knowledge.* London: Routledge.

Maykut, P., and R. Moorehouse. (1994). *Beginning qualitative research: A philosophical and practical guide.* London: Falmer.

Mayo, C. (1997). Foucauldian cautions on the subject and the educative implications of contingent identity. *Philosophy of Education Society Yearbook.* http://www.Ed.uiuc.edu/EPS/PES-Yearbook/97_docs/mayo.htm.

McDonough, K. (1993). Overcoming ambivalence about Foucault's relevance for education. *Philosophy of Education Society Yearbook.* http://www.Ed.uiuc.edu/EPS/PES-Yearbook/93_docs/MCDONOUG.htm.

McHoul, A., and W. Grace. (2000). *A Foucault primer: Discourse, power and the subject.* Carlton, AU: Melbourne University Press.

McIntosh, M (1992). Liberalism and contradictions of sexual politics. In L. Segal and M. McIntosh (Eds.), *Sex exposed: Sexuality and the pornography debate.* London: Virago.

McLaren, P. (1997). Critical pedagogy and the pragmatics of justice. In M. Peters (Ed.), *Education and the post-modern condition.* Westport, CT: Bergin and Garvey.

Measor, L., and P. Sykes. (1992). *Gender and schooling.* London: Cassell.

Mernissi, F. (1987). *Beyond the veil: Male-female dynamics in modern Muslim society.* Bloomington: Indiana University Press.

Metcalf, B. (1990). *Perfecting women: Maulana Ashraf Ali Thanawi's Bihishti zewar: A partial translation with commentary.* Berkeley: University of California Press.

Millett, K. (1973). *Sexual politics.* London: Virago.

Mirza, H. (1997a). Introduction: Mapping a genealogy of black British feminism. In H. Mirza (Ed.), *Black British feminism: A reader.* London: Routledge.

———. (Ed.). (1997b). *Black British feminism: A reader.* London: Routledge.

Mitchell, J. (1984). *Women: The longest revolution.* London: Virago.

Moghadam, V. (1994). Introduction: Women and identity politics in theoretical and comparative perspective. In V. Moghadam (Ed.), *Identity politics and women: Cultural reassertions and feminisms in international perspective.* Boulder, CO: Westview Press.

Mohanty, C. (1995). Feminist encounters: Locating the politics of experience. In L. Nicholson and S. Seidman (Eds.), *Social post-modernism: Beyond identity politics*. Cambridge: Cambridge University Press.

Mohanty, S. (1993). The epistemic status of cultural identity: On beloved and the postcolonial condition. *Cultural Critique* 24 (Spring): 41–80.

———. (1995). Colonial Legacies, multicultural futures: Relativism, objectivity, and the challenge of otherness. *PMLA* 110 (1): 108–18.

Molyneux, M. (1985) "Mobilisation Without Emancipation? Women's Interests, State and Revolution in Nicaragua." Feminist Studies 11 (2): 227–54.

Molyneux, M. (1979). Beyond the domestic labor debate. *New Left Review* 116.

———. (2001). *Women's movements in international perspective: Latin America and beyond*. Houndmills, UK: Palgrave.

Moore-Gilbert, B. (1997). *Postcolonial theory: Contexts, practices, policies*. London: Verso.

Morales, A., and R. Morales. (1986). *Getting home alive*. Ithaca, NY: Firebrand Books.

Mouffe, C. (1994). *Democratic pluralism: A critique of rationalist approach*. Toronto: Faculty of Law, University of Toronto.

———. (1995). Feminism, citizenship, and radical democratic politics. In L. Nicholson and S. Seidman (Eds.), *Social post-modernism: Beyond identity politics*. Cambridge: Cambridge University Press.

Moya, P. M. (1997). Postmodernism, "realism", and the politics of identity: Cherrie Moraga and Chicana feminism. In J. Alexander and C. Mohanty (Eds.), *Feminist genealogies, colonial legacies, democratic futures*. New York: Routledge.

Mumtaz, K. (1994). Identity politics and women: "Fundamentalism" and women in Pakistan. In V. Moghadam (Ed.), *Identity politics and women: Cultural reassertions and feminism in international perspective*. Boulder, CO: Westview.

———. (1996). The gender dimension in Sindh's ethnic conflict. In K. Rupesinghe and K. Mumtaz (Eds.), *Internal conflicts in South Asia*. London: Sage.

Mumtaz, K., and F. Shaheed. (1987). *Women of Pakistan: Two steps forward, one step back?* Lahore: Vanguard.

———. (1991). Historical roots of the women's movement: The period of awakening 1896–1947. In F. Zafar (Ed.). *Finding our way: Readings on women in Pakistan*. Lahore: ASR.

Naheed, K. (1986). *Siyah hashiyay main gulabi rang* (Red color in black borders, Urdu). Lahore: Sang e Meel Publications.

Nias, J. (1993). Primary teachers talking: A reflexive account of longitudinal research. In Martyn Hammersley (Ed.), *Educational research: Current issues*. London: Paul Chapman with the Open University.

Nandy, A. (1983). *The intimate enemy: Loss and recovery of self under colonialism*. New Delhi: Oxford University Press.

———. (2003). *The romance of the state and the fate of dissent in the tropics*. New Delhi: Oxford University Press.

Narayan, U. (1989). Perspectives from a non-western feminist. In A. Jagger and S. Bordo (Eds.), *Gender/body/knowledge*. New Brunswick, NJ: Rutgers University Press.

———. (1997). *Dislocating cultures: Identities, transitions and third world feminism*. New York: Routledge.

Naseem, M. A. (1995). Promoting the development of a culture of peace and devising innovative methods for early prevention and peaceful management of conflicts: A view from Pakistan. In *The culture of peace in Central South Asia*. Rawalpindi: UNESCO/Friends.

———. (2004), Gendered identity, and the educational discourse in Pakistan. In P. Ninnes and S. Mehta (Eds.). *Re-Imagining the discourse: Post foundational ideas for comparative education*. London: Routledge.

Nash, K. (1997). The feminist critique of liberal individualism. *Journal of Political Ideologies* 2 (1).

Nasr, S. (1992). Democracy and the crisis of governability. *Asian Survey* 32 (June): 521–37.

National College of Educational Research. (1987). *Urdu ki Nai Kitab*. New Delhi: Butra Art Press.

National Commission on the Status of Women in Pakistan. (1995). *Report*. Islamabad: Government of Pakistan.

Nayyar, A. (1998). Madrassah education frozen in time. In P. Hoodbhoy (Ed.), *Education and the state: Fifty years of Pakistan*. Karachi: Oxford University Press.

Nayyar, A., and A. Salim (n.d.) *The subtle subversion: The state of curricula and textbooks in Pakistan; Urdu, English, social studies and civics*. Islamabad: SDPI.

Nicholson, L. (Ed.). (1990). *Feminism/post-modernism*. London: Routledge.

———. (1995). Interpreting gender. In L. Nicholson and S. Seidman (Eds.), *Social post-modernism: Beyond identity politics*. Cambridge: Cambridge University Press.

Noman, O. (1990). *Pakistan: A political and economic history since 1947*. London: Kegan Paul International.

Okin, S. (1991). Gender, the public and the private. In D. Held (Ed.). *Political theory today*. London: Polity Press.

Oxfam (2002). *Education for all (EFA): Fast track or slow trickle?* Briefing Paper 43. http://www.oxfam.org/eng/pdfs/pp030408_educ_efa_wbimf.pdf.

Oxhorn, P. (1995). *Organizing civil society: The popular sectors and the struggle for democracy in Chile*. University Park: Pennsylvania State University Press.

Palermo, J. (2002). *Poststructuralist readings of the pedagogical encounter*. New York: Peter Lang.

Pandey, G. (1992). *The construction of communalism in colonial North India*. Delhi: Oxford University Press.

Pasha, M. K. (1997). The hyper-extended state: Civil society and democracy. In R. B. Rais (Ed.), *State, society and democratic change in Pakistan*. Karachi: Oxford University Press.

Pateman, C. (1988a). *The sexual contract*. London: Polity Press.

Pateman, C. (1988b). The patriarchal welfare state. In A. Guttman (ed.), *Democracy and the welfare state*. Princeton, NJ: Princeton University Press.

Petchesky, R. (1990). *Abortion and women's choice: The state, sexuality, and reproductive freedom*. Boston: Northeastern University Press.

Peters, M. (Ed.). (1999). *After the disciplines: The emergence of cultural studies*. Westport, CT: Bergin & Garvey.

Phelan, S. (1989). *Identity politics*. Philadelphia: Temple University Press.

Pillow, W. (1997). Decentring silences/troubling irony: Teen pregnancy's challenge to policy analysis. In C. Marshall (Ed.), *Feminist critical policy analysis I: A primary and secondary schooling perspective*. London: Falmer Press.

———. (2000). Exposed methodology: The body as a deconstructive practice. In E. St. Pierre and W. Pillow (Eds.), *Working the ruins: Feminist poststructural theory and methods in education*. New York: Routledge.

Popkewitz, T. (Ed.). (1987). *Critical studies in teacher education: Its folklore, theory and practice*. London: Falmer.

Prakash, G. (1992). Writing post-orientalist histories of the Third World: Indian historiography is good to think. In N. Dirks (Ed.), *Colonialism and culture*. Ann Arbor: University of Michigan Press.

Pringle, R., and S. Watson. (1992). Women's interests and the post-structuralist state. In M. Barrett and A. Phillips (eds.), *Destabilizing theory: Contemporary feminist debates*. Cambridge: Polity Press.

Punjab Textbook Board (PTB). (n.d.a.). *Meri Kitab* 1 (My book 1). Lahore: Khalid Book Center.

———. (n.d.b.). *Meri Kitab* 2 (My book 2). Lahore: Maktaba Tameer Insaniyat.

———. (n.d.c.). *Meri Kitab* 3 (My book 3). Lahore: Sindh Offset Printers and Publishers.

———. (n.d.d.). *Meri Kitab* 4 (My book 4). Lahore: Wisdom Publishing.

———. (n.d.e.). Urdu 8. Lahore: H.Y. Printers and Publishers.

———. (n.d.f.). Social studies 4. Lahore: Muhammadi Copy House.

———. (n.d.g.). Social studies 5. Lahore: Aziz Book Depot.

———. (n.d.h.). Social studies 6. Lahore: Lion Press Pvt.

———. (n.d.i.). Social studies 7. Lahore: Ideal Book House.

———. (n.d.j.). Social studies 8. Lahore: Makka Publishing House.

———. (1971). *Muasharati uloom* (Social studies) for class 3. Lahore: Sh. Barkat Ali and Sons.

———. (1971a). *Urdu ki chothi kitab* (for class 4). Lahore: Sh. Ghulam Ali and Sons.

———. (1972). *Urdu ki aathween kitab*. Lahore. Sh. Barkat Ali and Sons.

———. (1973). *Social studies* for class 3. Lahore: Sh. Barkat Ali and Sons.

———. (1973a). *Urdu ki aathween kitab*. Lahore. Kitabistan Publishing.

———. (1975). *Social studies: History and civics* for English Medium Schools for class 6. Lahore: Syedsons Publishers.

———. (1976). *Social studies* for class 5. Lahore: Izhar Sons.

———. (1976a). *Social studies* for class 7. Lahore: Maktaba-e-Mueen-ul-Adab.

———. (1978). *Social studies* (Sheikhupura District Supplement) for class 3. Lahore: Zafar Book Stall.

———. (1980). *Social studies* (Lahore District Supplement) for class 3. Lahore: Adara-e- Farough-e- Urdu.

————. (1980b). *Social studies* (Rawalpindi District Supplement) for class 3. Lahore: Modern Book Depot.

————. (1982). *Social studies* for class 5. Lahore: Sh. M. Ashraf.

————. (1984). *Muasharati uloom* (Social studies) for class 6. Lahore: New Crescent.

————. (1987). *Muasharati uloom* (Social studies) for class 7. Lahore: Khana-e-Anjuman-e-Himayat-e-Islam.

————. (1988). *Social studies* for class 5. Lahore: Khana-e-Anjuman-e-Himayat-e-Islam.

————. (1992). *Social studies* for class 7. Lahore: Madina Printing and Publishing House.

————. (1993). *Urdu ki chatti kitab* (for class 6). Lahore: Sartaj Book Depot.

————. (1993a). *Urdu ki panchween kitab* (for class 5). Lahore: Sartaj Book Depot.

————. (2002). *Meri kitab* for class 1 (Urdu). Lahore: Izhar Sons.

————. (2002a). *Meri kitab* for class 2 (Urdu). Lahore: Izhar Sons.

————. (2002b). *Meri kitab* for class 3 (Urdu). Lahore: Izhar Sons.

————. (2002c). *Urdu* for class 4. Lahore: Izhar Sons.

————. (2002d). *Urdu* for class 5. Lahore: Izhar Sons.

————. (2002e). *Urdu* for class 6. Lahore: Izhar Sons.

————. (2002f). *Urdu* for class 7. Lahore: Izhar Sons.

————. (2002g). *Urdu* for class 8. Lahore: Izhar Sons.

————. (2002h). *Social studies* for class 4. Lahore: Izhar Sons.

————. (2002i). *Social studies* for class 5. Lahore: Izhar Sons.

————. (2002j). *Social studies* for class 6. Lahore: Izhar Sons.

————. (2002k). *Social studies* for class 7. Lahore: Izhar Sons.

————. (2002l). *Social studies* for class 8. Lahore: Izhar Sons.

————. (2002m). *Muasharati uloom* (Urdu) for class 7. Lahore: Izhar Sons.

————. (2006). *Meri Kitab 5* (My book 5). Lahore: Maktaba Dastan.

————. (2009a). Urdu 6. Lahore: Kamran.

————. (2009b). Urdu 7. Lahore: West Pak .

Quddus, S. (1979). *Education and national reconstruction of Pakistan*. Lahore: S. I. Gillani.

Qureshi, I. (1975). *Education in Pakistan: An inquiry into objectives and achievements*. Karachi: Ma'aref Ltd.

Rabine, L. (1988). A feminist politics of non-identity. *Feminist Studies* 14 (1): 11–31.

Rabinow, P. (1984). *The Foucault reader*. New York: Pantheon Books.

Racine, J. (2002). Introduction: The state and nation of Pakistan—A multifaceted assessment. In S. Mumtaz, J. Racine, and I. Ali (Eds.), *Pakistan: The contours of state and society*. Karachi: Oxford University Press.

Rahman, F. (1953). *New education in the making of Pakistan*. London: Cassell.

Rahman, I. (1997). Enemy images on Pakistan television. In Z. Mian and I. Ahmad (Eds.), *Making enemies, creating conflict: Pakistan's crises of state and society*. Lahore: Mashal.

Rahman, T. (1998c). *Language and feminist issues in Pakistan*. Working Paper Series 31. Islamabad: SDPI.

Rai, S. (1996). Women and the state in the third world: Some issues for debate. In S. Rai and G. Lievesley (Eds.), *Women and the state: International perspectives*. London: Taylor and Francis.

Ramanzanoglu, C. (1989). *Feminism and the contradictions of oppression*. London: Routledge.

Randal, V. (1987). *Women and politics*. Basingstoke, UK: Macmillan.

———. (1988). Gender and power: Women engage the state. In V. Randal and G. Waylen (Eds.), *Gender, politics and the state*. New York: Routledge.

Ray, R., and A. Korteweg. (1999). Women's movements in the third world: Identity, mobilization and autonomy. *Annual Review of Sociology* 25: 47–71.

Reskin, B., and H. Hartmann. (Eds.) (1981). *Women's work, men's work: Sex segregation on the job*. Committee on Women's Employment and Related Social Issues. Commission on Behavioral and Social Sciences and Education. National Research Council. Washington, DC: National Academy Press.

Richardson, D. (1997). Deconstructing feminist critiques of radical feminism. In M. Ang-Lydgate, C. Corrin, and M. Henry (Eds.), *Desperately seeking sisterhood: Still challenging and building*. London: Taylor and Francis.

Richter, W. I. (1986). The political meaning of Islamization in Pakistan: Prognosis, implications, and questions. In A. Weiss (Ed.), *Islamic reassertion in Pakistan: The application of Islamic laws in a modern state*. Syracuse, NY: Syracuse University Press.

Rizvi, H. A. (2000). *The military and politics in Pakistan, 1947–86*. Lahore: Progressive Publishers.

Roberts, K. (1998). *Deepening democracy? The modern left and social movements in Chile and Peru*. Stanford, CA: Stanford University Press.

Rorty, R. (1989). *Contingency, irony and solidarity*. Cambridge: Cambridge University Press.

———. (1996). Solidarity or objectivity. In Lawrence E. Cahoone (Ed.), *From modernism to Post-modernism: An anthology*.

Rousse, S. (1994). Gender(ed) struggles: The state, religion and society. In K. Bhasin, R. Menon, and N. Khan (Eds.), *Against all odds: Essays on women, religion and development from India and Pakistan*. New Delhi: Kali for Women.

———. (1998). The outsider(s) within: Sovereignty and citizenship in Pakistan. In P. Jeffery and A. Basu, (Eds.), *Appropriating gender: Women's activism and politicized religion in South Asia*. New York: Routledge.

Rowbotham, S. (1973). *Hidden from history: 300 years of women's oppression and the fight against it*. London: Pluto Press.

———. (1992). *Women in movement: Feminism and social action*. London: Routledge.

Rutherford, J. (1990). A place called home: Identity and the cultural politics of difference. In J. Rutherford (Ed.), *Identity: Community, culture, difference*. London: Lawrence and Wishart.

Saeed, H., and A. Khan. (2000). Legalized cruelty: Anti women laws in Pakistan. In J. Mirsky and M. Radlett (Eds.), *No paradise yet: The world's women face the new century*. London: Panos.

Said, E. (1978). *Orientalism*. London: Routledge and Kegan Paul.

————. (1994). *Culture and imperialism*. New York: Vintage.

Saigol, R. (1993). *Education: Critical perspectives*. Lahore: Progressive.

————. (1994). Domestic knowledge systems and patriarchy. In N. Khan, A. Zia, and R. Saigol, *Unveiling the issues: Pakistani women's perspectives on social, political and ideological issues*. Transcribed and translated by Naureen Amjad and Rubina Saigol. Lahore: ASR.

————. (1994a). The sharia bill and its impact on women. In K. Bhasin, R. Menon, and N. Khan (Eds.), *Against all odds: Essays on women, religion and development from India and Pakistan*. New Delhi: Kali for Women.

————. (1995). *Knowledge and identity: Articulation of gender in educational discourse in Pakistan*. Lahore: ASR.

————. (1997). Militarization, nation and gender. In Z. Mian and I. Ahmad (Eds.), *Making enemies, creating conflict: Pakistan's crises of state and society*. Lahore: Mashal.

Saqib, G. (1983). *Modernization of muslim education in Egypt, Pakistan and Turkey: A comparative study*. Lahore: Islamic Book Service.

Sarup, M. (1989). *An introductory guide to post-structuralism and post-modernism*. Athens: University of Georgia Press.

Sathar, Z. (1996). Women's schooling and autonomy as factors in fertility change in Pakistan: Some empirical evidence. In R. Jeffery and A. Basu (Eds.). *Girls schooling, women's autonomy, and fertility change in South Asia*. New Delhi: Sage.

Sawicki, J. (1988). Identity politics and sexual freedom: Foucault and feminism. In I. Diamond and L. Quinby (Eds.), *Feminism and Foucault: Reflections on resistance*. Boston: Northeastern University Press.

Sayeed, K. (1967). *The political system of Pakistan*. Boston: Houghton Mifflin.

————. (1980). *Politics in Pakistan: The nature and direction of change*. New York: Praeger.

Scheurich, J. (1997). *Research method in the postmodern*. London: Falmer.

Schor, N. (1989). This existentialism which is not one: Coming to grips with Irigaray. *Differences* 1 (2): 38–58.

Scott, J. (1986). Gender: A useful category of historical analysis. *American Historical Review* 91 (5): 1053–75.

————. (1988). *Gender and the politics of history*. New York: Columbia University Press.

Sedgwick, E. (1985). *Between men: English literature and male homosexual desire*. New York: Columbia University Press.

Seidman, I. (1991). *Interviewing as qualitative research: A guide for researchers in education and the social sciences*. New York: Teachers College, Columbia University.

Seidman, S. (1992). *Embattled eros: Sexual politics and ethics in contemporary America*. New York: Routledge.

————. (1994). Symposium: Queer theory/sociology: A dialogue. *Sociological Theory* 12 (2): 166–248.

Shafqat, S. (2002). Democracy and political transformation in Pakistan. In S. Mumtaz, J. Racine, and I. Ali (Eds.), *Pakistan: The contours of state and society*. Karachi: Oxford University Press.

Shah Nawaz, M. (1990). *The heart divided*. Lahore: ASR.

Shah, S. M. (1994). *Federalism in Pakistan: Theory and practice*. Islamabad: Chair on Quaid-i-Azam and Freedom Movement, National Institute of Pakistan Studies, Quaid-i-Azam University.

———. (1996). *Religion and politics in Pakistan: 1972–88*. Islamabad: Quaid-i-Azam Chair, National Institute of Pakistan Studies, Quaid-i-Azam University.

Shaheed, F. (1990). *Pakistan's women: An analytical description*. Lahore: Sanjh.

———. (1991). The cultural articulation of patriarchy. In F. Zafar (Ed.), *Finding our way: Readings on women in Pakistan*. Lahore: ASR.

———. (1998). The other side of the discourse: Women's experiences of identity, religion and activism in Pakistan. In P. Jeffery and A. Basu, (Eds.), *Appropriating gender: Women's activism and politicized religion in South Asia*. New York: Routledge.

Shahnawaz, J. (1971). *Father and daughter: A political autobiography*. Lahore: Oxford.

Showstock-Sassoon, A. (Ed.). (1987). *Women and the state: Shifting the boundaries of the public and the private*. London: Hutchinson.

Sindh Textbook Board. (2001). *Asaan Urdu* for class 4. Karachi: Ilm o Aql Book Depot.

———. (2001a). *Urdu*. Karachi: Sh. Shaukat Ali and Sons.

Skocpol, T. (1985). Bringing the state back in: Strategies of analysis in current research. In P. Evans, D. Ruechemeyer, and T. Skocpol (Eds.), *Bringing the state back in*. Cambridge: Cambridge University Press.

Slater, D. (Ed.) (1985). *New social movements and the state in Latin America*. Amsterdam.

Spivak, G. (1990). *The post-colonial critic: Interviews, strategies, dialogues*. Sara Harasym (Ed.). New York: Routledge.

———. (1993). Can the subaltern speak? In P. Williams and L. Chrisman (Eds.), *Colonial discourse and the postcolonial theory*. Hemel Hempstead, UK: Harvester Wheatsheaf.

———. (1996). Poststructuralism, marginality, postcoloniality and value. In P. Mongia (Ed.), *Contemporary postcolonial theory*. London: Arnold.

St. Pierre, E. (1997). An introduction to figuration—A poststructural practice of inquiry. *International Journal of Qualitative Studies in Education* 10 (3): 279–84.

St. Pierre, E., and W. Pillow. (Eds.). (2000). *Working the ruins: Feminist poststructural theory and methods in education*. New York: Routeledge.

Stacey (1986). Are feminists afraid to leave home? The challenge of conservative pro-family feminism. In J. Mitchell and A. Oakley (Eds.), *What is feminism*. Oxford: Blackwell.

Stephano, C. Di (1990). Dilemmas of difference: Feminism, modernity and postmodernism. In L. Nicholson (Ed.), *Feminism/post-modernism*. New York: Routledge.

Stetson, D., and A. Mazur. (Eds.). (1995). *Comparative state feminism*. London: Sage.

Syed, A. (1978). Z. A. Bhutto's self-characterization and Pakistani political culture. *Asian Survey* 18 (12): 1250–66.

Tajfel, H., and J. Turner. (1986). The social identity theory of intergroup behavior. In S. Worchel and W. Austin (Eds.), *Psychology of intergroup relations.* Chicago: Nelson-Hall.

Taseer, S. (1980), *Bhutto: A political biography.* Ithaca, NY: Cornell University Press.

Taylor, C. (1988). *The postmodern position.* http://www.sil.org.

———. (1989). *Sources of the self.* Cambridge, MA: Harvard University Press.

———. (1992). *Multiculturalism and the politics of recognition.* Princeton, NJ: Princeton University Press.

Taylor, S. (2004) "Gender Equity and Education: What Are the Issues Now?' In Bruce Burnett, Daphne Meadmore and Gordon Tait (Eds.), *New Questions for Contemporary Teachers.* French's Forest, NSW: Pearson Education Australia, pp. 87–100.

Tong, R. (1989) *Feminist thought: A comprehensive introduction.* Sydney: Unwin and Allen.

Treiman, D., and H. Hartmann. (Eds.). (1981). *Women, work, wages: Equal pay for jobs of equal values.* Committee on Occupational Classification and Analysis. Assembly of Behavioral and Social Sciences. National Research Council. Washington, DC: National Academy Press.

UNDP (2001). *Portrayal of women in media.* Islamabad: UNDP.

UNESCO (2002). *The future of girls' education in Pakistan: A study on policy measures and other factors determining girls' education.* A report by Dr. H. Khalid and Dr. E Mukhtar. Islamabad. UNESCO.

Venkataramani, M. (1982). *The American role in Pakistan: 1947–1958.* New Delhi: Radiant.

Villenas, S. (1996). The colonizer/colonized Chicana ethnographer: Identity, marginalization, and co-optation in the field. *Harvard Educational Review* 66: 711–31.

Vogel, L. (1995). *Woman questions: Essays for a materialist feminism.* London: Pluto Press.

Walby, S. (1990). *Theorizing patriarchy.* Oxford: Blackwell.

———. (Ed.). (1994). Is citizenship gendered? *Sociology* 28 (2): 379–95.

Walkerdine, V. (1984). Development psychology and the child centered pedagogy. In J. Henriques et al. *Changing the subject.* London: Methuen.

———. (1986). Post-structuralist theory and everyday social practice: The family and the school. In S. Wilkinson (Ed.). *Feminist social psychology: Developing theory and Practice.* Milton Keynes, UK: Open University Press.

Waseem, M. (1987). *Pakistan under martial law: 1977–1985.* Lahore: Vanguard.

———. (1994). *Politics and the state in Pakistan.* Islamabad: National Institute of Historical and Cultural Research.

———. (2001). Underdevelopment of social sciences in Pakistan. In S. Hashmi (Ed.), *The state of social sciences in Pakistan.* Islamabad: Quaid-i-Azam University. http://www.coss.sdnpk.org/chap11.htm.

Watson, S. (Ed.) (1990). *Playing the state: Australian feminist intervention.* London: Verso.

———. (1990a). The state of play: An introduction. In S. Watson (Ed.), *Playing the state: Australian feminist intervention.* London: Verso.

Watson, S., and R. Pringle. (1990). Fathers, brothers, mates: The fraternal state in Australia. In S. Watson (Ed.). *Playing the state: Australian feminist intervention.* London, New York: Verso.

Waylen, G. (1996). Democratization, feminism and the state: The establishment of SERNAM in Chile. In S. Rai and G. Lievesley (Eds.). *Women and the state: International perspectives.* London: Taylor and Francis.

Weedon, C. (1987). *Feminist practice and post-structuralist theory.* Oxford: Blackwell.

Weiss, A. (Ed.). (1986). *Islamic reassertion in Pakistan: The application of Islamic laws in a modern state.* Syracuse, NY: Syracuse University Press.

———. (1992). *Walls within walls: Life histories of working women in the old city of Lahore.* Boulder, CO: Westview Press.

———. (1994). The consequences of state policy for women in Pakistan. In M. Weiner and A. Banuazizi (Eds.), *The politics of social transformation in Afghanistan, Iran and Pakistan.* Syracuse, NY: Syracuse University Press.

———. (1998). The slow yet steady path to women's empowerment in Pakistan. In Y. Haddad and J. L. Esposito (Eds.), *Islam, gender and social change.* New York: Oxford University Press.

West, C. (1993). *Beyond eurocentrism and multiculturalism.* Monroe, ME: Common Courage Press.

West Pakistan Textbook Board. (1964). *Urdu ka qaida* for class 1. Lahore.

———. (1969). *Muasharati Uloom.* Lahore.

Williams, P., and L. Chrisman. (Eds.). (1995). *Colonial discourse and post-colonial theory: A reader.* New York: Columbia University Press.

Williams, R. (1977). Structures of feeling. In *Marxism and Literature.* Oxford: Oxford University Press.

Wittgenstein, L. (1972). *Philosophical investigations.* Translated by G. E. M. Anscombe. Oxford: Basil Blackwell.

Wolf, N. (1993). *Fire with fire: The new female power and how it will change the 21st Century.* London: Chatto and Windus.

Wood, B. (1998). Stuart Hall's cultural studies and the problem of hegemony. *British Journal of Sociology* 49 (3): 399–414.

Young, I. (1995). Gender as seriality: Thinking about women as a social collective. In L. Nicholson and S. Seidman (Eds.), *Social post-modernism: Beyond identity politics.* Cambridge: Cambridge University Press.

Young, S. (1997). *Changing the wor(l)d: Discourse, politics and the feminist movement.* New York: Routledge.

———. (1988). Gender and politics in Pakistan. In R. Sharma (Ed.), *Representation of gender, democracy and identity politics in relation to South Asia.* New Delhi: Indian Books.

———. (1991). Women's education: Problems and prospects. In F. Zafar (Ed.), *Finding our way: Readings on women in Pakistan.* Lahore: ASR.

Zia, A. (1994). Women and media: An overview. In N. S Khan, A. Zia, and R. Saigol (Eds.), *Unveiling the issues: Pakistani women's perspectives on social, political and ideological issues.* Transcribed and translated by Naureen Amjad an Rubina Saigol. Lahore: ASR.

Zia, S. (1991). The legal status of women in Pakistan. In F. Zafar (Ed.), *Finding our way: Readings on women in Pakistan.* Lahore: ASR.

# Index